PRAISE FOR THE *MANGA GUIDE* SERIES

"Highly recommended."
—CHOICE MAGAZINE ON *THE MANGA GUIDE TO DATABASES*

"Stimulus for the next generation of scientists."
—SCIENTIFIC COMPUTING ON *THE MANGA GUIDE TO MOLECULAR BIOLOGY*

"A great fit of form and subject. Recommended."
—OTAKU USA MAGAZINE ON *THE MANGA GUIDE TO PHYSICS*

"The art is charming and the humor engaging. A fun and fairly painless lesson on what many consider to be a less-than-thrilling subject."
—SCHOOL LIBRARY JOURNAL ON *THE MANGA GUIDE TO STATISTICS*

"This is really what a good math text should be like. Unlike the majority of books on subjects like statistics, it doesn't just present the material as a dry series of pointless-seeming formulas. It presents statistics as something *fun*, and something enlightening."
—GOOD MATH, BAD MATH ON *THE MANGA GUIDE TO STATISTICS*

"I found the cartoon approach of this book so compelling and its story so endearing that I recommend that every teacher of introductory physics, in both high school and college, consider using it."
—AMERICAN JOURNAL OF PHYSICS ON *THE MANGA GUIDE TO PHYSICS*

"The series is consistently good. A great way to introduce kids to the wonder and vastness of the cosmos."
—DISCOVERY.COM ON *THE MANGA GUIDE TO THE UNIVERSE*

"A single tortured cry will escape the lips of every thirty-something biochem major who sees *The Manga Guide to Molecular Biology*: 'Why, oh why couldn't this have been written when I was in college?'"
—THE SAN FRANCISCO EXAMINER

"Scientifically solid . . . entertainingly bizarre."
—CHAD ORZEL, AUTHOR OF *HOW TO TEACH PHYSICS TO YOUR DOG*, ON *THE MANGA GUIDE TO RELATIVITY*

"A lot of fun to read. The interactions between the characters are lighthearted, and the whole setting has a sort of quirkiness about it that makes you keep reading just for the joy of it."
—HACK A DAY ON *THE MANGA GUIDE TO ELECTRICITY*

"*The Manga Guide to Databases* was the most enjoyable tech book I've ever read."
—RIKKI KITE, LINUX PRO MAGAZINE

"The *Manga Guides* definitely have a place on my bookshelf."
—SMITHSONIAN'S "SURPRISING SCIENCE"

"For parents trying to give their kids an edge or just for kids with a curiosity about their electronics, *The Manga Guide to Electricity* should definitely be on their bookshelves."
—SACRAMENTO BOOK REVIEW

"This is a solid book and I wish there were more like it in the IT world."
—SLASHDOT ON *THE MANGA GUIDE TO DATABASES*

"*The Manga Guide to Electricity* makes accessible a very intimidating subject, letting the reader have fun while still delivering the goods."
—GEEKDAD BLOG, WIRED.COM

"If you want to introduce a subject that kids wouldn't normally be very interested in, give it an amusing storyline and wrap it in cartoons."
—MAKE ON *THE MANGA GUIDE TO STATISTICS*

"A clever blend that makes relativity easier to think about—even if you're no Einstein."
—STARDATE, UNIVERSITY OF TEXAS, ON *THE MANGA GUIDE TO RELATIVITY*

"This book does exactly what it is supposed to: offer a fun, interesting way to learn calculus concepts that would otherwise be extremely bland to memorize."
—DAILY TECH ON *THE MANGA GUIDE TO CALCULUS*

"The art is fantastic, and the teaching method is both fun and educational."
—ACTIVE ANIME ON *THE MANGA GUIDE TO PHYSICS*

"An awfully fun, highly educational read."
—FRAZZLEDDAD ON *THE MANGA GUIDE TO PHYSICS*

"Makes it possible for a 10-year-old to develop a decent working knowledge of a subject that sends most college students running for the hills."
—SKEPTICBLOG ON *THE MANGA GUIDE TO MOLECULAR BIOLOGY*

"This book is by far the best book I have read on the subject. I think this book absolutely rocks and recommend it to anyone working with or just interested in databases."
—GEEK AT LARGE ON *THE MANGA GUIDE TO DATABASES*

"The book purposefully departs from a traditional physics textbook and it does it very well."
—DR. MARINA MILNER-BOLOTIN, RYERSON UNIVERSITY ON *THE MANGA GUIDE TO PHYSICS*

"Kids would be, I think, much more likely to actually pick this up and find out if they are interested in statistics as opposed to a regular textbook."
—GEEK BOOK ON *THE MANGA GUIDE TO STATISTICS*

THE MANGA GUIDE™ TO BIOCHEMISTRY

THE MANGA GUIDE™ TO
BIOCHEMISTRY

MASAHARU TAKEMURA,
KIKUYARO, AND
OFFICE SAWA

The Manga Guide to Biochemistry is a translation of the Japanese original, *Manga de wakaru seikagaku*, published by Ohmsha, Ltd. of Tokyo, Japan, © 2009 by Masaharu Takemura and Office Sawa

This English edition is co-published by No Starch Press, Inc. and Ohmsha, Ltd.

20 19 18 17 4 5 6 7 8 9

ISBN-10: 1-59327-276-6
ISBN-13: 978-1-59327-276-0

Publisher: William Pollock
Author: Masaharu Takemura
Illustrator: Kikuyaro
Producer: Office Sawa
Production Editor: Serena Yang
Developmental Editors: Keith Fancher and Sondra Silverhawk
Translator: Arnie Rusoff
Technical Reviewers: Brandon Budde and Jordan Gallinetti
Compositor: Riley Hoffman
Copyeditor: Kristina Potts
Proofreader: Alison Law
Indexer: BIM Indexing & Proofreading Services

For information on book distributors or translations, please contact No Starch Press, Inc. directly:
No Starch Press, Inc.
245 8th Street, San Francisco, CA 94103
phone: 1.415.863.9900; info@nostarch.com; http://www.nostarch.com/

Library of Congress Cataloging-in-Publication Data
A catalog record of this book is available from the Library of Congress.

CONTENTS

4
ENZYMES ARE THE KEYS TO CHEMICAL REACTIONS 149

5
MOLECULAR BIOLOGY AND THE BIOCHEMISTRY OF NUCLEIC ACIDS 199

PREFACE

This book introduces the world of biochemistry in an approachable comic format.

Biochemistry is a synthesis of biology and chemistry, which together elucidate the processes of life at the most basic level. It is the study of the molecules that constitute our bodies and those of other living organisms, and the chemical reactions that occur within cells. In recent years, the field of biochemistry has been growing at an unprecedented rate. From the end of the 19th century and into the 20th, scientists have conducted chemical research on phenomena in the fields of medicine, nutritional science, agriculture, biology, and many other subjects, and this research has led to some incredible discoveries.

When you consider the diversity of the fields listed above, biochemistry may seem like a disjointed collection of different sciences. But even though the objectives differ, the concepts on which they are based are the same—the chemical elucidation of life phenomena. Therefore, the fundamentals of biochemistry must be learned by anyone who intends to participate in any field that deals with the human body or life phenomena to any extent, such as medicine, dentistry, pharmacology, agriculture, nutritional science, and nursing.

This book explains the most important points in biochemistry in an easy-to-understand format. It can be used as a reference book or supplementary reader for a biochemistry course, or for a course in medical science or nutritional science. You can use this book as a quick refresher or to gain a better understanding of this fascinating science. Even a high school student would certainly be able to comprehend this material.

The organization of this book differs somewhat from other existing biochemistry books. For example, although the major cellular chemical components (substances that are present in all living things: saccharides, lipids, nucleic acids, and proteins) are usually described first in an ordinary biochemistry textbook, discussions of each of these substances are incorporated organically, rather than in an independent chapter. I did this because I believe that introducing these substances in context makes them easier to understand and remember.

In addition, I've included information about biochemistry in our everyday lives in Chapter 3 to highlight the significance of biochemistry by showing how it applies to subjects that most people are familiar with.

The protagonist of this book is a high school girl named Kumi who is very concerned with dieting. I chose this story because it relates to my own educational background as a member of a nutritional science division in an agricultural sciences department. These days, when people talk about biochemistry, the discussion often centers around nutrition and health. Many people are concerned with the phenomena that make up *metabolic syndrome*, a general name for the risk factors of an increasingly-common collection of disorders: type 2 diabetes, coronary artery disease, and stroke.

When I was writing this book, I had the entire text checked at both the manuscript and scenario stages by Professor Yukio Furuichi (emeritus professor at Mie University and currently a professor at Nagoya Women's University), whose specialty is lipid biochemistry, and Professor Shonen Yoshida (emeritus professor at Nagoya University and currently a consultant at the Cancer Immunotherapy Center of Nagoya Kyoritsu Hospital), whose specialties are biochemistry and molecular biology. Professor Furuichi provided guidance for my graduate thesis, and Professor Yoshida provided guidance for my PhD thesis. I would like to take this opportunity to express my deep gratitude to both of them for taking time from their busy schedules to proofread this manuscript.

I would also like to take this opportunity to thank Professor Kazuo Kamemura, my mentor during my graduate student days, and his graduate student, Mitsutaka Ogawa, both of Nagahama Institute of Bio-Science and Technology. Specifically, I would like to thank them for the lectin blotting data that they provided. I would also like to thank everyone at the Ohmsha Development Bureau for their ongoing help on my previous work, *The Manga Guide to Molecular Biology*; Sawako Sawada of Office Sawa; the manga artist Kikuyaro, who created the delightful scenario and drawings; and, above all, you for choosing to read this book.

MASAHARU TAKEMURA
JANUARY 2009

I'M HOME!
OH, GOOD, YOU'RE BACK. HEY–
WAIT A SEC! I'LL BE RIGHT BACK!
AAARGH! I STILL HAVEN'T LOST ANY WEIGHT!
GOAL: LOSE 5 LBS!
DOWN WITH THE POUNDS!

I'VE GOT TO GET TO A HEALTHY WEIGHT!
バタ
バタ
UM... HELLO?
HUH?!
I JUST DROPPED BY TO OFFER YOU SOME FRUIT FROM MY GARDEN, BUT-
EEEEK!!
NEMOTO? WHERE DID YOU COME FROM?!
キャーーッ

WELL...
I'VE GOT TO HAND IT TO YOU, NEMOTO...

THIS IS ONE SCRUMPTIOUS MELON!

BUT I'M ON A DIET AND SHOULDN'T BE EATING FRUIT.
ずーん
I MAY HAVE SCREWED UP BIG TIME...

THERE'S NO REASON TO FEEL THAT WAY, KUMI.

I THINK YOU COULD EAT WHATEVER YOU WANTED, AND YOU'D STILL LOOK REALLY... UM...GREAT.
もじ
もじ
YEAH RIGHT! MY ENTIRE BODY IS PROBABLY MADE OF PIZZA AND CAKE!
(KUMI'S FAVORITE FOODS)

THAT DOES IT! I'M GOING TO FAST UNTIL I REACH MY GOAL!
I REFUSE TO BE OVERWEIGHT FOR EVEN ONE MORE DAY!

BUT KUMI...
THAT IS RIDICULOUS.
キラッ

YOU'VE GOT IT ALL WRONG!
FIRST OF ALL,
YOU'RE NOT OVERWEIGHT, AND...
...YOU'RE ALREADY ATTRACTIVE, AND...UM...
?
BLUSH

IN ANY CASE, YOU DON'T SEEM TO UNDERSTAND HOW THE HUMAN BODY WORKS!
AHEM
I'M ACTUALLY RESEARCHING THIS KIND OF THING AT MY UNIVERSITY.
やさしい
生化学*
* BIOCHEMISTRY 101
BIO... BIO-WHAT?
IT'S BIOCHEMISTRY!

IT LOOKS TOO HARD... I DON'T THINK I'D BE ABLE TO FOLLOW.
LET'S START WITH SOMETHING FAMILIAR THEN.

CALORIES, FAT, AND CARBOHYDRATES...
YOU KNOW WHAT *THOSE* ARE, RIGHT?

OF COURSE I DO! I'M ON A DIET, AFTER ALL.

NOW CHECK THIS OUT!

SERIOUSLY, TAKE A LOOK, OKAY?
DIETING: A SPECIAL REPORT
GETTING A SLIM SUMMER BODY
SO, FAT IS AN EXAMPLE OF A HIGH-CALORIE NUTRIENT, RIGHT?
SAYING THAT CARBOHYDRATES ARE HIGH IN CALORIES IS A LITTLE DIFFERENT, BUT PEOPLE OFTEN SAY THAT YOU'LL GET FAT IF YOU EAT TOO MANY CARBS.

CARBS
FAT
DESSERT
OBVIOUSLY! I ALREADY KNEW THAT!
WELL, I DON'T KNOW *WHY*...BUT MAGAZINES DON'T LIE, RIGHT?
UMM...

GAINING WEIGHT MEANS THAT FAT BUILDS UP IN YOUR BODY, RIGHT?
WHY DO YOU THINK YOU GAIN WEIGHT IF YOU EAT TOO MANY CARBOHYDRATES?

* ABOUT THE AUTHOR

THIS PROFESSOR IS...
SO BEAUTIFUL!!!
やさしい
生化学
CHEMISTRY ISN'T AS DIFFICULT AS YOU THINK, KUMI.
FOR EXAMPLE, WHEN YOU EAT DINNER AND DIGEST YOUR FOOD, THAT'S A CHEMICAL REACTION.
WHAT? NO WAY!
SO CHEMICAL REACTIONS MUST BE HAPPENING IN OUR BODIES ALL THE TIME, RIGHT?
THAT'S CORRECT!
OUR BODIES (AND THOSE OF OTHER LIVING CREATURES) ARE ACTUALLY MADE UP OF MANY TYPES OF CHEMICALS.
PROTEINS
WATER
VITAMINS
CARBOHYDRATES
FAT
MINERALS
THE FAT AND CARBOHYDRATES THAT WE TALKED ABOUT EARLIER ARE ALSO CHEMICALS, RIGHT?
THEY'RE ALL CHEMICALS!

I WAS SO BUSY WORRYING ABOUT MY WEIGHT AS A NUMBER...
I WASN'T THINKING ABOUT MY BODY FROM A CHEMICAL POINT OF VIEW.

EXACTLY!
TO SUM IT UP:
BIOCHEMISTRY IS THE STUDY OF WHAT'S OCCURRING INSIDE OUR BODIES (AND THE BODIES OF OTHER LIVING ORGANISMS)...
...WITH A SPECIAL FOCUS ON THIS "CHEMICAL POINT OF VIEW."
I'M ACTUALLY PERFORMING RESEARCH ON THE BODY'S CHEMICAL PROCESSES AT MY SCHOOL.
IF YOU WANT TO, YOU CAN JOIN ME AT THE LABORATORY FOR AN EXPERIMENT.
もじ もじ
UM...BY THE WAY...
THE NEXT DAY—
IF I PARTICIPATE IN AN EXPERIMENT, I CAN MEET THAT PROFESSOR!
SURE! I'LL DO IT!
AND I'LL LOOK LIKE A SUPERMODEL IN NO TIME!
*東西南北大学
* KREBS UNIVERSITY

黒坂研究室*
*KUROSAKA LABS
HELLO.
NICE TO MEET YOU, KUMI.
WELCOME TO MY LAB.
SHE'S EVEN MORE GORGEOUS IN PERSON!!!
?
UM...I'VE REALLY WANTED TO ASK YOU...

IF I STUDY BIOCHEMISTRY... WILL I BECOME AS BEAUTIFUL AS YOU?!
WHEN I SAW YOUR PICTURE, I WAS TOTALLY SMITTEN!
SWOON ♥
OH MY!
WELL, BIOCHEMISTRY AND YOUR PHYSICAL APPEARANCE AREN'T DIRECTLY RELATED...
BUT BIOCHEMISTRY CERTAINLY DEEPENS OUR UNDERSTANDING OF THE WAY IN WHICH OUR BODIES INTERACT WITH FOOD.

WE CAN STUDY THE WAY OUR BODIES CHEMICALLY BREAK DOWN WHAT WE EAT AND HOW IT'S TRANSFORMED INTO NUTRIENTS THAT THE BODY USES TO REPLENISH ITSELF.
THIS KNOWLEDGE CAN ALSO HELP US CURE DISEASES...
AND PROMOTE GOOD GENERAL HEALTH.
IF YOU TRULY UNDERSTAND HOW YOUR BODY WORKS...
YOU WILL BE HEALTHY AND BEAUTIFUL!
COOOOOL!
IF I STUDY BIOCHEMISTRY...
I MIGHT BECOME BEAUTIFUL LIKE THE PROFESSOR!
AND I CAN UNLOCK THE SECRETS OF HEALTH!!
I'LL DO IT!
THAT'S THE SPIRIT!
OKAY...
FIRST YOU NEED TO DRINK THIS WATER.

IT CONTAINS A ROBOT SO TINY THAT IT CAN'T BE SEEN WITH THE NAKED EYE.
WE'LL BE USING IT TO OBSERVE THE INSIDE OF YOUR BODY.
ROBOT & MASCOT
NICKNAME: ROBOCAT
DEVELOPED BY KUROSAKA LABS
DOWN THE HATCH!
GLUG GLUG GLUG
EEP!
NOW, AT LONG LAST...
LET THE STUDY OF BIOCHEMISTRY BEGIN!

WHAT HAPPENS INSIDE YOUR BODY?

1. Cell Structure

YOU LEARNED ABOUT CELLS IN BIOLOGY CLASS, RIGHT?
YEP!
CELLS ARE LIKE TINY POUCHES THAT MAKE UP OUR BODIES!
AMOEBAE, BACTERIA, AND OTHER TINY ORGANISMS ARE "UNICELLULAR MICROORGANISMS," WHICH MEANS THEY'RE MADE UP OF A SINGLE CELL.
AMOEBA
BACTERIA
LIVING CREATURES THAT ARE VISIBLE TO THE NAKED EYE—LIKE HUMANS, DOGS, OR PLANTS—ARE "MULTICELLULAR ORGANISMS," AND THEY'RE MADE UP OF MANY CELLS!
THAT'S RIGHT!
FOR EXAMPLE, A SINGLE ADULT BODY CONSISTS OF AN UNBELIEVABLY LARGE NUMBER OF CELLS... BETWEEN 60 AND 100 TRILLION.
THE CELL IS THE SMALLEST UNIT INSIDE OUR BODIES THAT CAN BE CLASSIFIED AS "LIVING."
I BET EVEN THE PROFESSOR'S CELLS ARE BEAUTIFUL!
TEE HEE
HEY, THE IMAGE FINISHED DOWNLOADING!
TAP
TAP
TAP
EEEEEEK!!! THAT'S TOTALLY GROSS...
KUMI'S CELLS
PRETTY SNAZZY, EH?
LET'S TRY ZOOMING IN ON A SINGLE CELL.
ZOOM
TAP

WHAT ARE THE COMPONENTS OF A CELL?

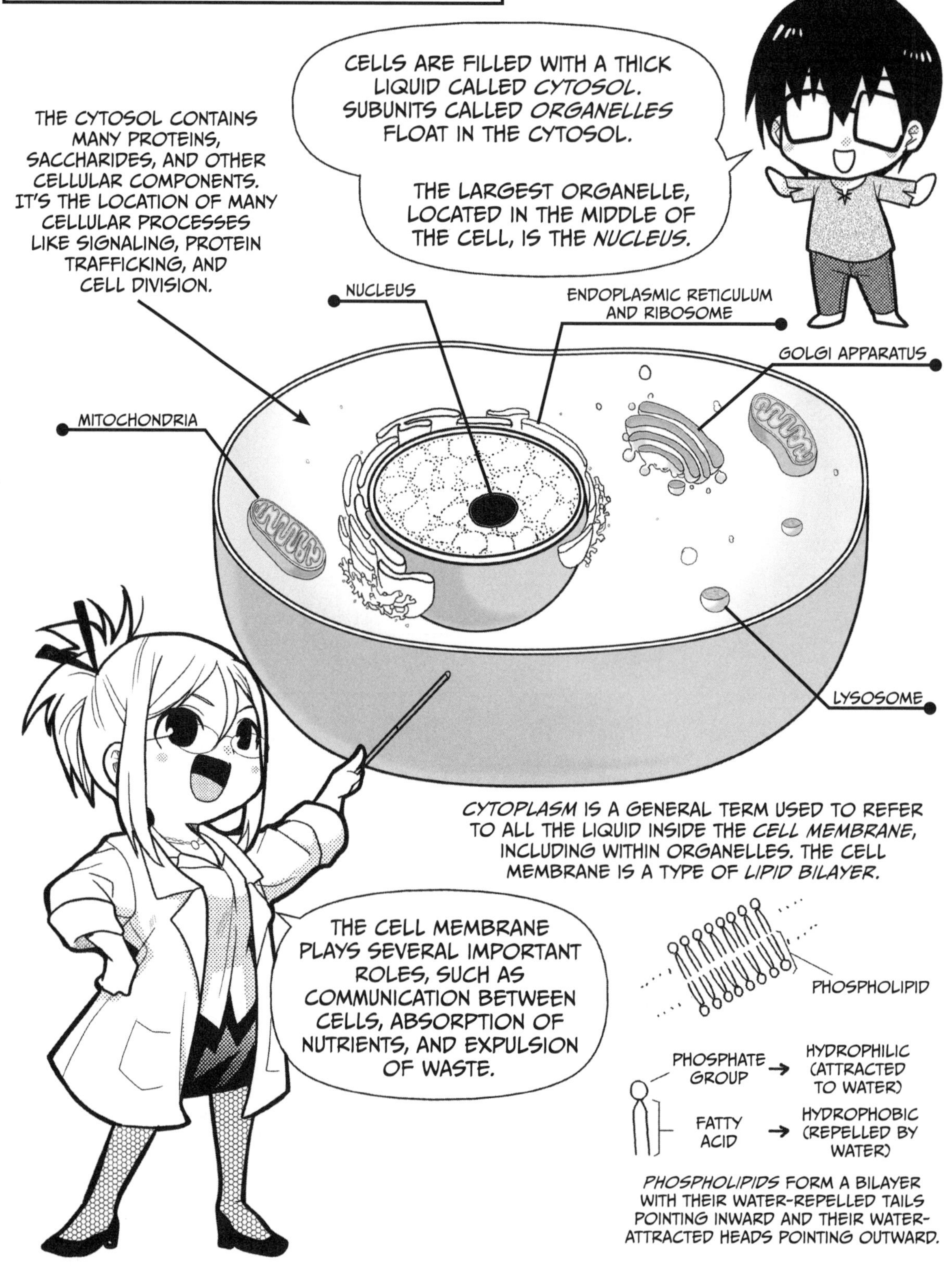

CELLS ARE FILLED WITH A THICK LIQUID CALLED *CYTOSOL*. SUBUNITS CALLED *ORGANELLES* FLOAT IN THE CYTOSOL.

THE LARGEST ORGANELLE, LOCATED IN THE MIDDLE OF THE CELL, IS THE *NUCLEUS*.

THE CYTOSOL CONTAINS MANY PROTEINS, SACCHARIDES, AND OTHER CELLULAR COMPONENTS. IT'S THE LOCATION OF MANY CELLULAR PROCESSES LIKE SIGNALING, PROTEIN TRAFFICKING, AND CELL DIVISION.

NUCLEUS

ENDOPLASMIC RETICULUM AND RIBOSOME

GOLGI APPARATUS

MITOCHONDRIA

LYSOSOME

CYTOPLASM IS A GENERAL TERM USED TO REFER TO ALL THE LIQUID INSIDE THE *CELL MEMBRANE*, INCLUDING WITHIN ORGANELLES. THE CELL MEMBRANE IS A TYPE OF *LIPID BILAYER*.

THE CELL MEMBRANE PLAYS SEVERAL IMPORTANT ROLES, SUCH AS COMMUNICATION BETWEEN CELLS, ABSORPTION OF NUTRIENTS, AND EXPULSION OF WASTE.

PHOSPHOLIPID

PHOSPHATE GROUP → HYDROPHILIC (ATTRACTED TO WATER)

FATTY ACID → HYDROPHOBIC (REPELLED BY WATER)

PHOSPHOLIPIDS FORM A BILAYER WITH THEIR WATER-REPELLED TAILS POINTING INWARD AND THEIR WATER-ATTRACTED HEADS POINTING OUTWARD.

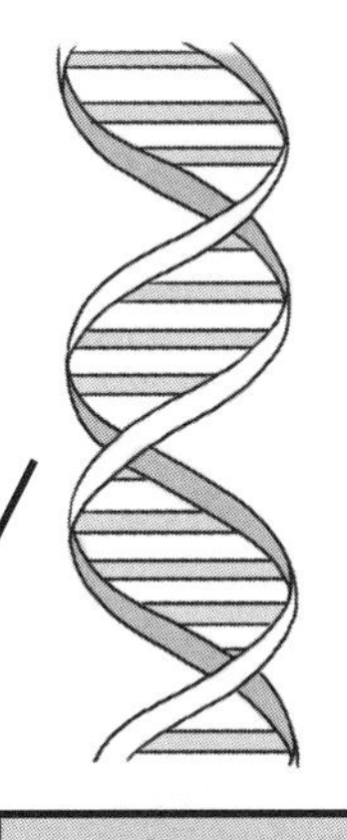

THE NUCLEUS CONTAINS *DEOXYRIBONUCLEIC ACID*, OR *DNA*, WHICH ENCODES GENES AND IS SOMETIMES REFERRED TO AS THE "BLUEPRINT" FOR LIFE.

THE NUCLEUS IS REFERRED TO AS THE "CONTROL CENTER" OF THE CELL.

NUCLEUS

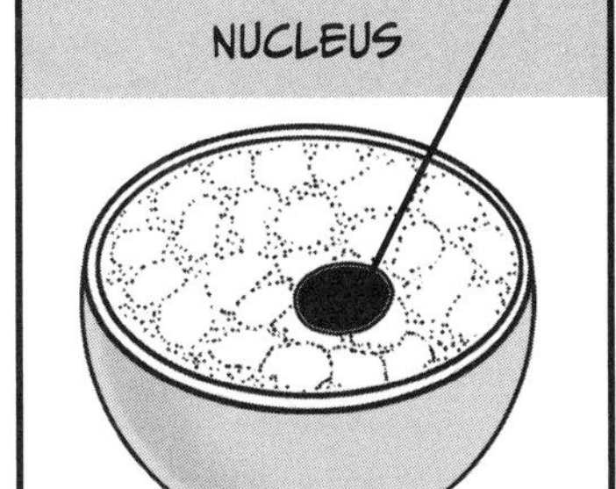

MITOCHONDRIA

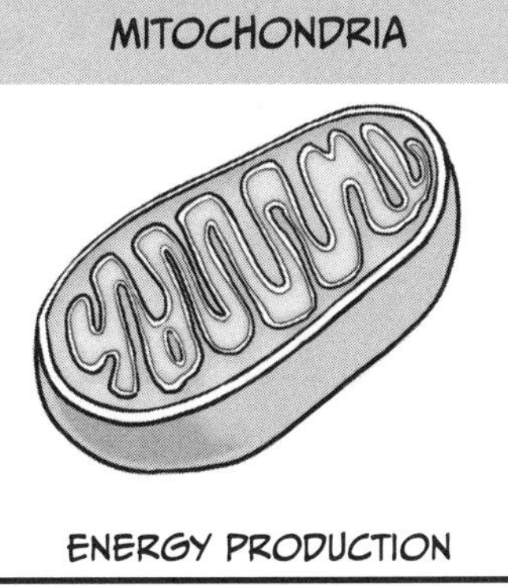

ENERGY PRODUCTION

ENDOPLASMIC RETICULUM AND RIBOSOME

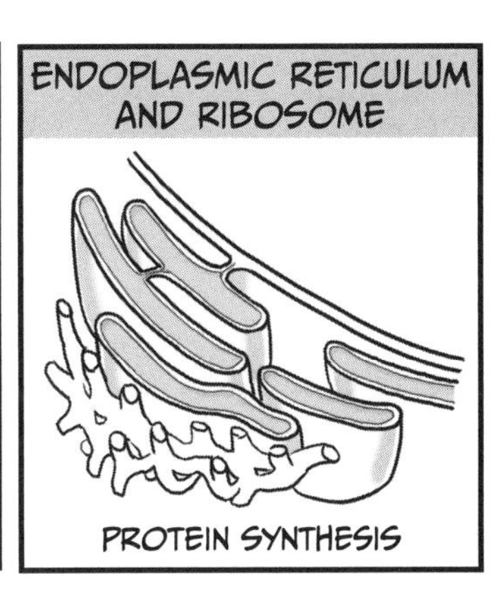

PROTEIN SYNTHESIS

GOLGI APPARATUS

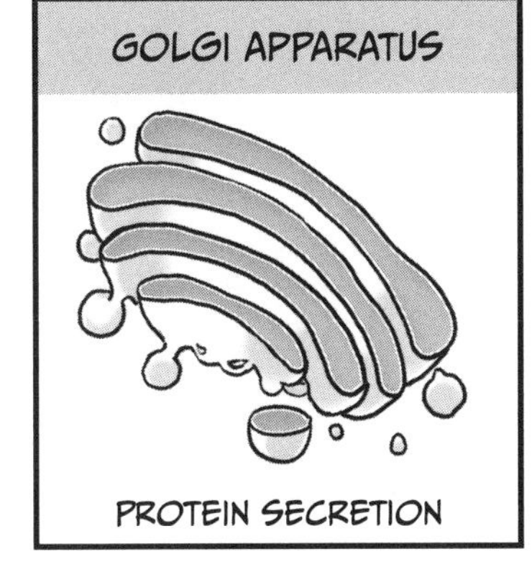

PROTEIN SECRETION

LYSOSOME

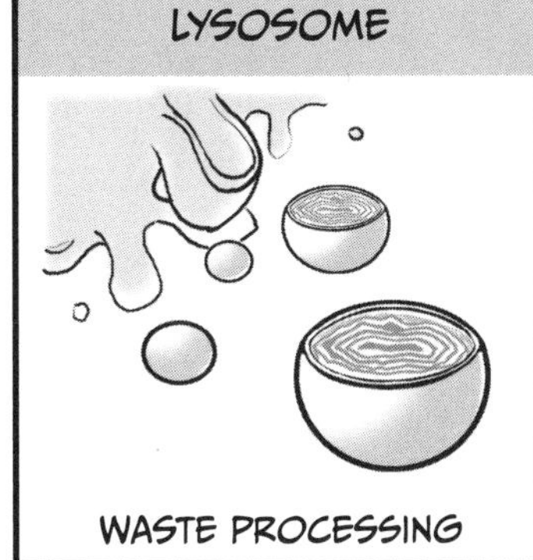

WASTE PROCESSING

CHLOROPLAST

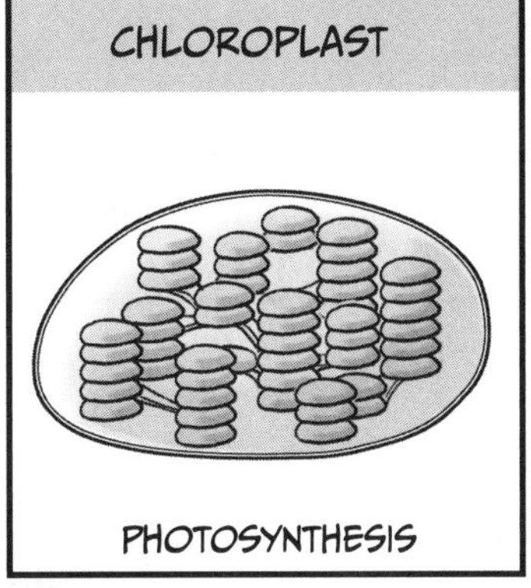

PHOTOSYNTHESIS

CHLOROPLASTS ARE FOUND ONLY IN PLANTS AND SOME MICROBES.

2. What Happens Inside a Cell?

PROTEIN SYNTHESIS

WHEN YOU HEAR "PROTEIN," YOU PROBABLY THINK OF THE NUTRIENTS FOUND IN FOODS, BUT...

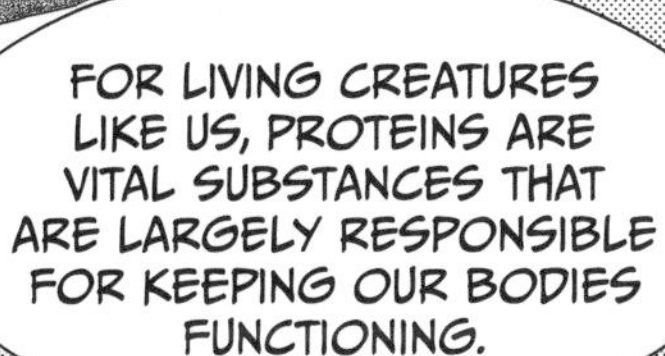

FOR LIVING CREATURES LIKE US, PROTEINS ARE VITAL SUBSTANCES THAT ARE LARGELY RESPONSIBLE FOR KEEPING OUR BODIES FUNCTIONING.

WOW, ARE PROTEINS REALLY THAT DELICIOUS, ER, I MEAN, IMPORTANT?

ABSOLUTELY! OUR BODIES ARE MAINTAINED BY DIFFERENT PROTEINS CARRYING OUT THEIR DUTIES.

- MAINTENANCE OF CELLULAR STRUCTURE
- DIGESTION
- MUSCLE CREATION
- PROTECTION FROM VIRAL, FUNGAL, AND PARASITIC INFECTIONS

PROTEINS ARE CONTINUOUSLY MANUFACTURED BY EVERY CELL IN OUR BODY.

REMEMBER WHEN ROBOCAT LOOKED AT THE DNA INSIDE THE NUCLEUS?

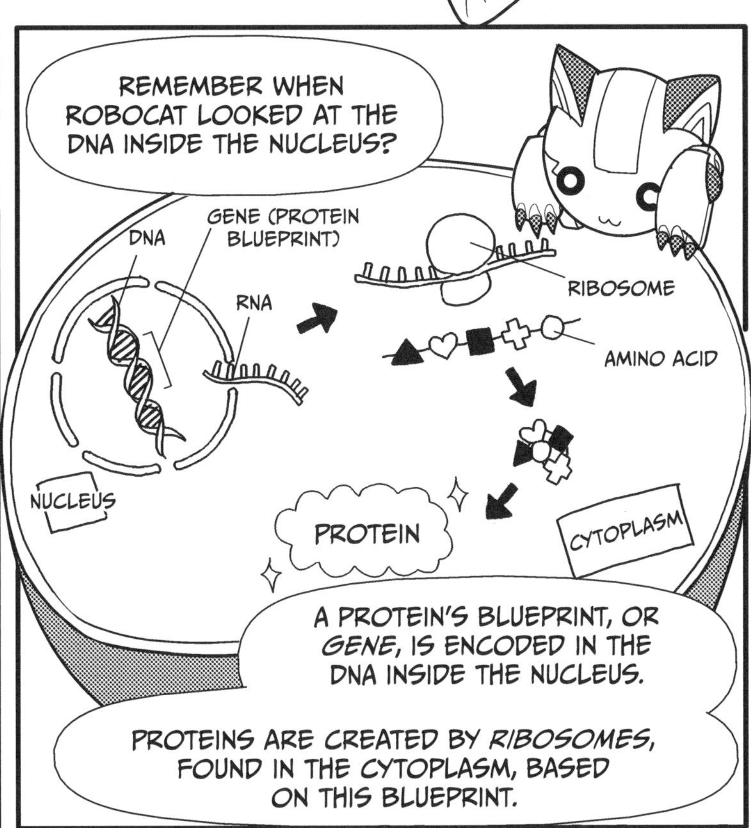

A PROTEIN'S BLUEPRINT, OR *GENE*, IS ENCODED IN THE DNA INSIDE THE NUCLEUS.

PROTEINS ARE CREATED BY *RIBOSOMES*, FOUND IN THE CYTOPLASM, BASED ON THIS BLUEPRINT.

THE RIBOSOMES ARE LIKE CHEFS FOLLOWING A RECIPE TO MAKE A MEAL!

METABOLISM

ONCE PROTEINS ARE CREATED, THEY DO IMPORTANT JOBS INSIDE AND OUTSIDE THE CELLS.

ONE OF THESE JOBS IS...

...CATALYZING THE BREAKDOWN OF FOODS OR MEDICINES THAT ENTER THE BODY INTO SOMETHING USEFUL

AND BREAKING DOWN UNNECESSARY OR HARMFUL SUBSTANCES INTO SOMETHING THAT CAN BE EXPELLED MORE EASILY.

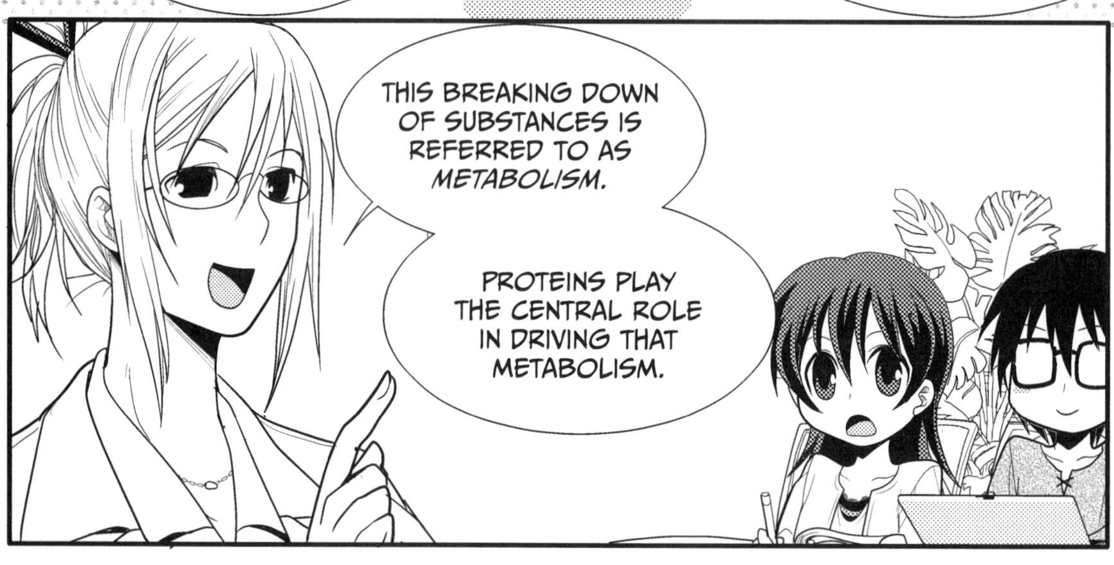

BREAKING DOWN FOOD INTO NUTRIENTS, ABSORBING THESE NUTRIENTS, AND CHANGING THEM INTO SUBSTANCES YOUR BODY CAN USE TO REPLENISH ITSELF... THESE ARE ALL JOBS FOR SPECIALIZED PROTEINS!

FOR EXAMPLE, SINCE ALCOHOL IS HIGHLY TOXIC TO THE BODY, IT'S BROKEN DOWN BY LIVER CELLS AND CHANGED INTO A NONTOXIC SUBSTANCE.

THIS IS ALSO THE JOB OF A SPECIALIZED PROTEIN!

THE MEDICINE YOU TAKE WHEN YOU'RE SICK NEEDS TO BE BROKEN DOWN AS WELL. PROTEINS IN THE LIVER HELP YOUR BODY SIMPLIFY THAT MEDICINE INTO SUBSTANCES THAT PRODUCE THE DESIRED HEALING EFFECT IN THE RIGHT LOCATION.

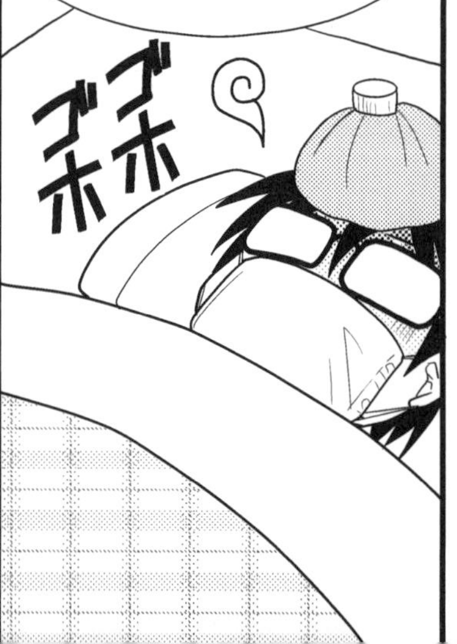

(PROTEINS, FATS, CARBOHYDRATES, VITAMINS, MINERALS, AND SO ON)
NUTRIENT
METABOLISM
BECOMES MATERIALS USABLE BY THE BODY
CREATES ENERGY
ALCOHOL
METABOLISM
DETOXIFIED
THINGS YOU EAT OR DRINK ARE GENERALLY METABOLIZED LIKE THIS.
I SEE...
FOR INSTANCE, HERE'S WHAT HAPPENS AFTER I DRINK A DELICIOUS GLASS OF WINE.
ALCOHOL PASSES THROUGH THE BLOOD AND INTO THE LIVER.
METABOLISM
METABOLISM
THE LIVER METABOLIZES THE ALCOHOL.
DETOXIFICATION
CARBON DIOXIDE
WATER
THE PROFESSOR SURE CAN HOLD HER LIQUOR...
METABOLISM IS PERFORMED BY PROTEINS.
IN THE CELL MEMBRANE, THE CYTOPLASM, THE NUCLEUS, AND EVERY OTHER ORGANELLE, THE ROLES ARE DIVIDED AMONG MANY PROTEINS SO THAT METABOLISM IS CONSTANTLY PERFORMED.
PROTEIN
METABOLISM
PROTEIN
METABOLISM
PROTEIN
METABOLISM
PROTEIN
METABOLISM
WOW! PROTEINS ARE DILIGENTLY WORKING AWAY INSIDE MY BODY EVEN WHEN I'M EATING DINNER OR SLEEPING OFF A COLD...
JEEZ, MY CELLS WORK HARDER THAN I DO...

ENERGY PRODUCTION
BY THE WAY, NEMOTO...
YOU SAID YOU WERE RUNNING LOW ON CASH THIS MONTH, DIDN'T YOU?
HEE HEE HEE
WHAT?!
NO, I'M NOT! I MEAN...
WELL, TO BE HONEST, I GUESS I AM A LITTLE STRAPPED FOR CASH.
IN TODAY'S MODERN SOCIETY, MONEY IS ESSENTIAL FOR ALMOST ANY ACTIVITY, RIGHT?
I GOTTA STOP TELLING MY PROFESSORS ABOUT MY MONEY PROBLEMS...
IN A SIMILAR WAY, CELLS HAVE SOMETHING THAT IS ESSENTIAL FOR THEIR ACTIVITY.
IT'S LIKE CURRENCY, BUT IT'S USED FOR CHEMICAL REACTIONS IN OUR CELLS.
ATP
THIS IS THE SUBSTANCE CALLED *ADENOSINE TRIPHOSPHATE*, OR *ATP*.
A-DEEN-OH... WHAT?
UM...LET'S JUST STICK WITH *ATP*.

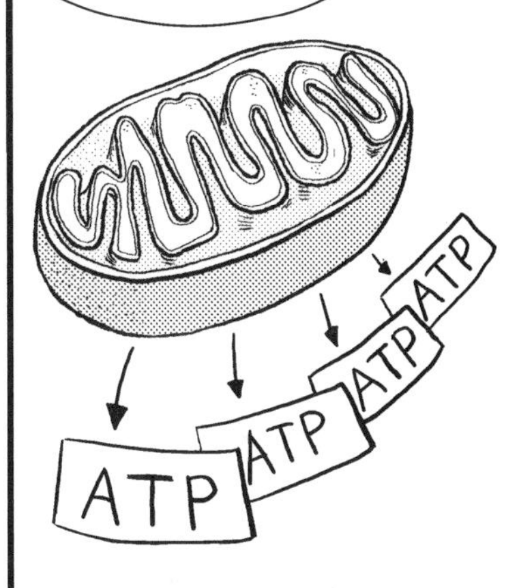

* THESE SACCHARIDES ARE ALSO KNOWN AS CARBOHYDRATES.

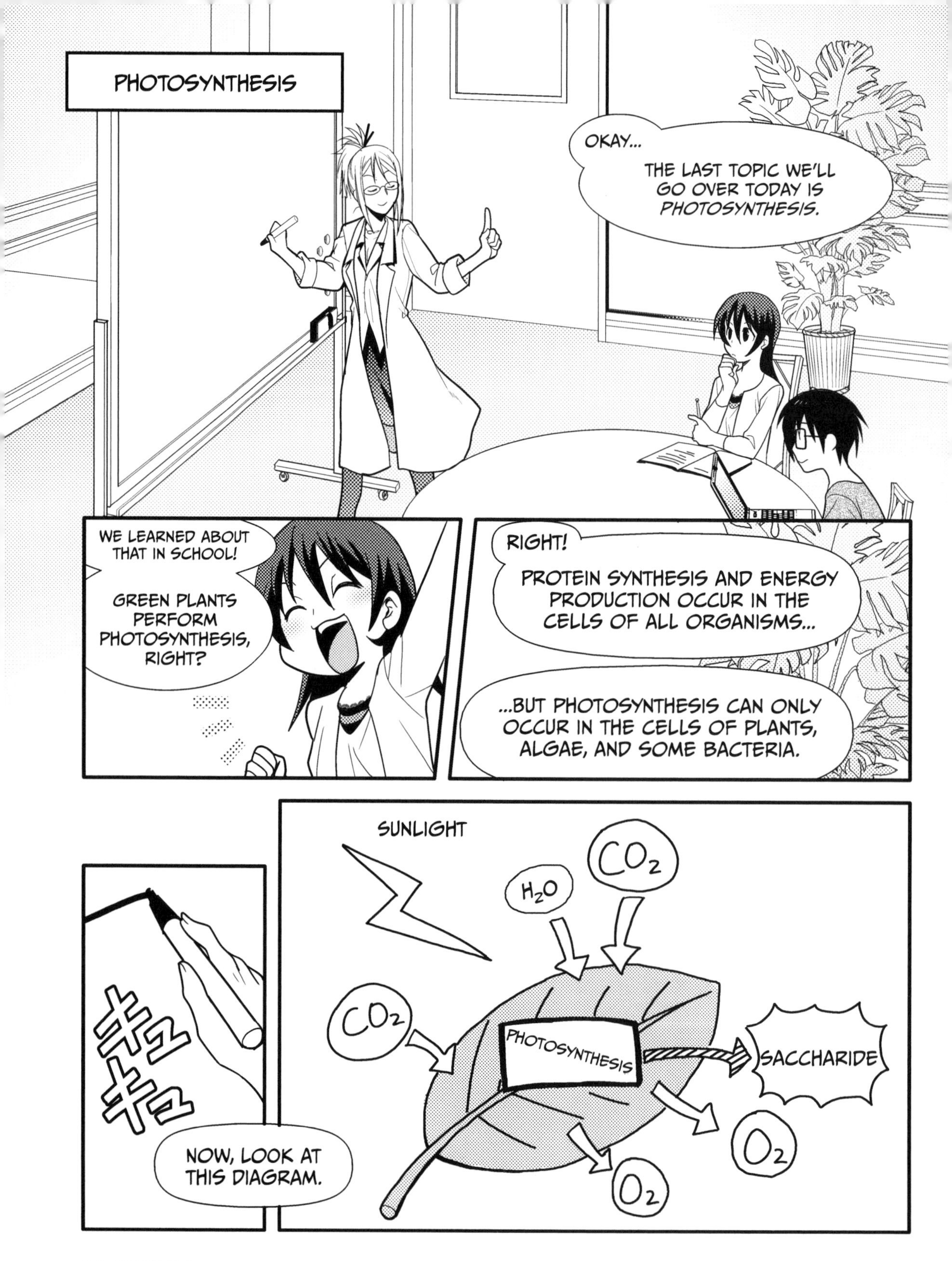
PHOTOSYNTHESIS
OKAY...
THE LAST TOPIC WE'LL GO OVER TODAY IS *PHOTOSYNTHESIS.*
WE LEARNED ABOUT THAT IN SCHOOL!
GREEN PLANTS PERFORM PHOTOSYNTHESIS, RIGHT?
RIGHT!
PROTEIN SYNTHESIS AND ENERGY PRODUCTION OCCUR IN THE CELLS OF ALL ORGANISMS...
...BUT PHOTOSYNTHESIS CAN ONLY OCCUR IN THE CELLS OF PLANTS, ALGAE, AND SOME BACTERIA.
キュキュ
NOW, LOOK AT THIS DIAGRAM.
SUNLIGHT
H_2O
CO_2
CO_2
PHOTOSYNTHESIS
SACCHARIDE
O_2
O_2

H₂O
CO₂
PHOTOSYNTHESIS
SACCHARIDE
PHOTOSYNTHESIS IS A REACTION THAT USES SUNLIGHT AND CARBON DIOXIDE TO SYNTHESIZE SACCHARIDES.
SNAP
SACCHARIDES WERE REQUIRED TO CREATE ATP, RIGHT?
AND OXYGEN IS CREATED AS A BY-PRODUCT OF PHOTOSYNTHESIS.

SO DO YOU UNDERSTAND WHY PLANTS ARE SO IMPORTANT FOR LIVING CREATURES LIKE US?
SACCHARIDES AND OXYGEN ARE REQUIRED TO CREATE ATP, WHICH IS ESSENTIAL FOR OUR BODIES...
AH!
AND BOTH OF THOSE THINGS ARE PRODUCED BY PHOTOSYNTHESIS!
I TOTALLY GET IT!

IF PLANTS DIDN'T PERFORM PHOTOSYNTHESIS, LIFE WOULD BE SO CRUEL.
UH, KUMI...
I HAVE NO ATP.
LITTLE MATCH GIRL KUMI

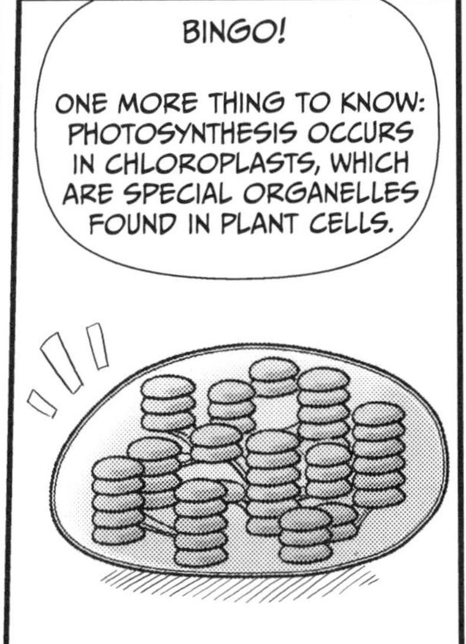
BINGO!
ONE MORE THING TO KNOW: PHOTOSYNTHESIS OCCURS IN CHLOROPLASTS, WHICH ARE SPECIAL ORGANELLES FOUND IN PLANT CELLS.

3. A Cell Is the Location of Many Chemical Reactions

BIOCHEMISTRY OF PROTEIN SYNTHESIS
WHAT DO YOU THINK HAPPENS WHEN PROTEINS ARE SYNTHESIZED?
AMINO ACIDS
ARE JOINED TOGETHER
THEN FOLDED
TO CREATE A PROTEIN!
A PROTEIN IS ACTUALLY FORMED BY MANY SMALL MOLECULES CALLED AMINO ACIDS JOINING TOGETHER.
THERE ARE 20 COMMON TYPES OF AMINO ACIDS USED TO CREATE PROTEINS.
PROTEINS
→ MUSCLE CONTRACTION (ACTIN AND MYOSIN)
→ ENZYMES
→ ANTIBODIES
→ HAIR (KERATIN)
→ SKIN (COLLAGEN)
AMINO ACIDS
THESE 20 TYPES CAN BE COMBINED IN DIFFERENT NUMBERS AND ORDERS TO CREATE VARIOUS TYPES OF PROTEINS.
THEY'RE LIKE LITTLE CANDY NECKLACES? HOW CUTE! AND HOW DELICIOUS...
PROTEIN SYNTHESIS IS CARRIED OUT BY RIBOSOMES, WHICH FLOAT IN THE CYTOPLASM OR ARE STUCK TO THE ENDOPLASMIC RETICULUM.
ZOOM!
RIBOSOME
A SNOWMAN?
ALTHOUGH THEY LOOK LIKE LITTLE GRAINS OF RICE, IF WE ZOOM IN WE CAN SEE THAT THEY HAVE A STRANGE SHAPE.
ACTUALLY, IF WE SIMPLIFY IT A BIT, A RIBOSOME LOOKS A LOT LIKE A SNOWMAN.

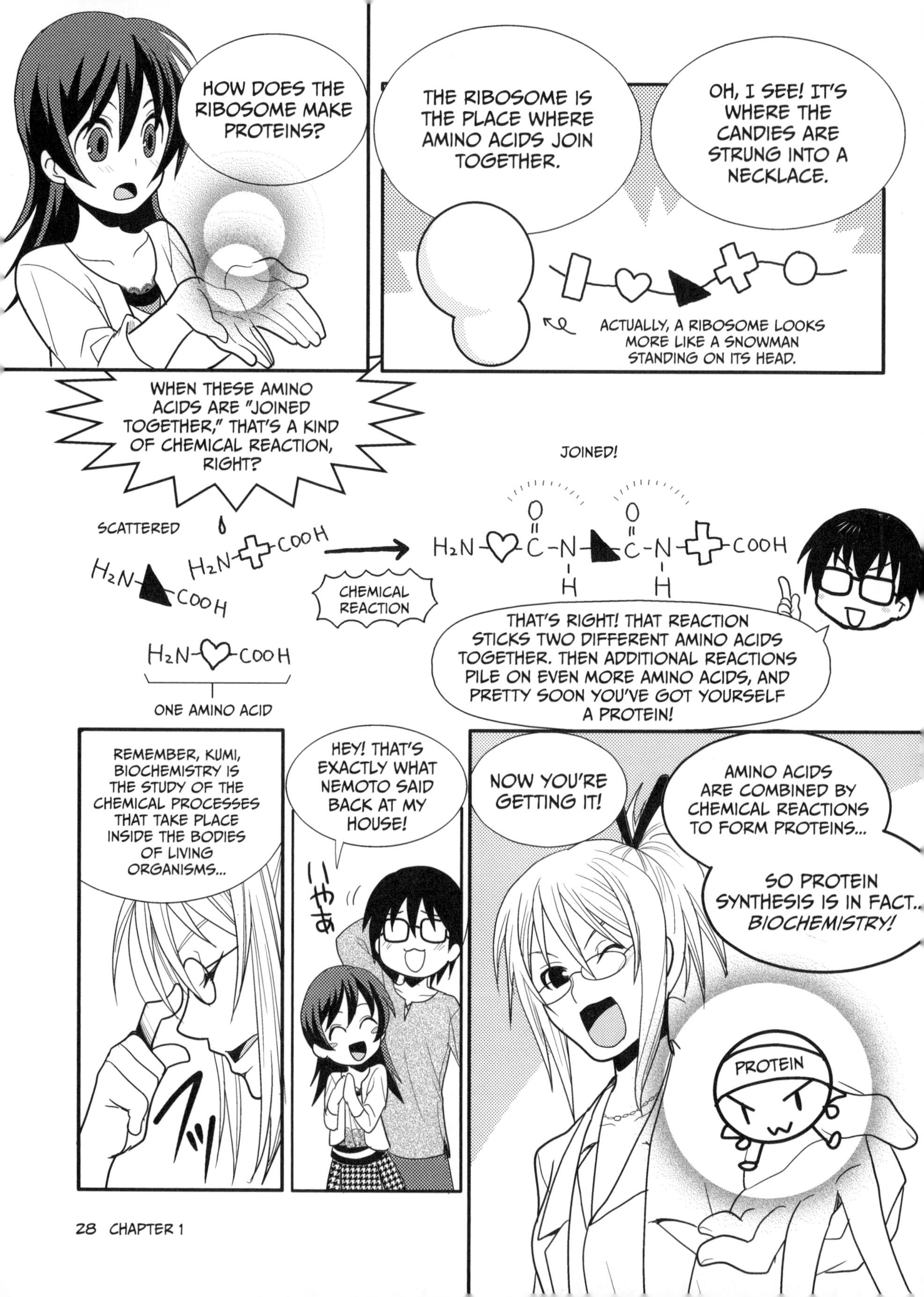
HOW DOES THE RIBOSOME MAKE PROTEINS?
THE RIBOSOME IS THE PLACE WHERE AMINO ACIDS JOIN TOGETHER.
OH, I SEE! IT'S WHERE THE CANDIES ARE STRUNG INTO A NECKLACE.
ACTUALLY, A RIBOSOME LOOKS MORE LIKE A SNOWMAN STANDING ON ITS HEAD.
WHEN THESE AMINO ACIDS ARE "JOINED TOGETHER," THAT'S A KIND OF CHEMICAL REACTION, RIGHT?
SCATTERED
H_2N-COOH
H_2N-COOH
H_2N-COOH
ONE AMINO ACID
CHEMICAL REACTION
JOINED!
H_2N-C(=O)-N(H)-C(=O)-N(H)-COOH
THAT'S RIGHT! THAT REACTION STICKS TWO DIFFERENT AMINO ACIDS TOGETHER. THEN ADDITIONAL REACTIONS PILE ON EVEN MORE AMINO ACIDS, AND PRETTY SOON YOU'VE GOT YOURSELF A PROTEIN!
REMEMBER, KUMI, BIOCHEMISTRY IS THE STUDY OF THE CHEMICAL PROCESSES THAT TAKE PLACE INSIDE THE BODIES OF LIVING ORGANISMS...
HEY! THAT'S EXACTLY WHAT NEMOTO SAID BACK AT MY HOUSE!
いやあ
NOW YOU'RE GETTING IT!
AMINO ACIDS ARE COMBINED BY CHEMICAL REACTIONS TO FORM PROTEINS...
SO PROTEIN SYNTHESIS IS IN FACT... *BIOCHEMISTRY!*
PROTEIN

BIOCHEMISTRY OF METABOLISM

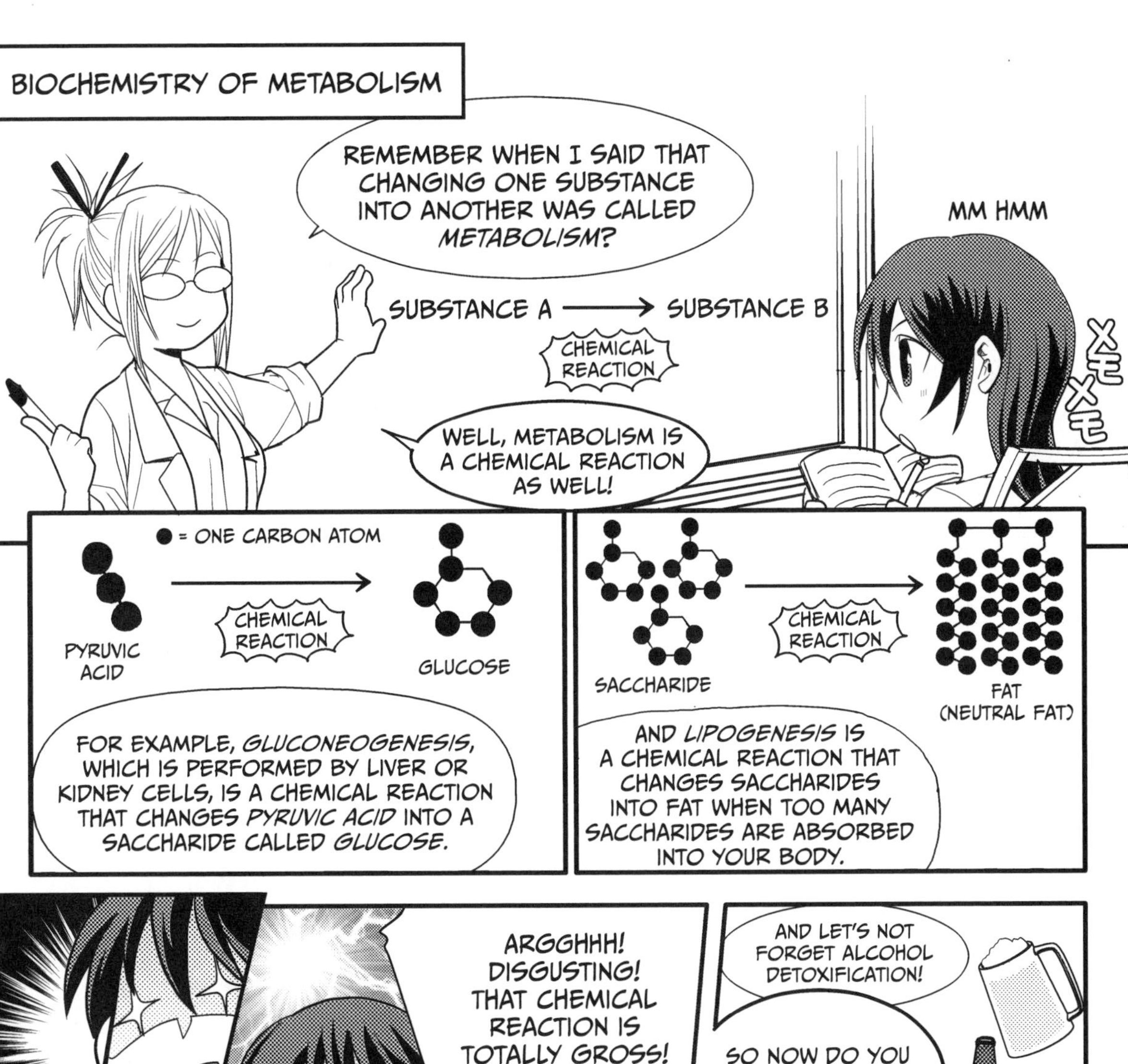

BIOCHEMISTRY OF ENERGY PRODUCTION
ENERGY PRODUCTION IS ALSO A KIND OF METABOLISM.
GLUCOSE
?
PYRUVIC ACID
SUBSTANCE A
SUBSTANCE B
CHEMICAL REACTION
TO PRODUCE ENERGY, GLUCOSE IS FIRST BROKEN DOWN INTO PYRUVIC ACID IN CYTOSOL.
HUH? DIDN'T YOU MENTION GLUCOSE AND PYRUVIC ACID EARLIER?
YUP! THIS IS THE REVERSE VERSION OF GLUCONEOGENESIS, CALLED GLYCOLYSIS.
GLYCOLYSIS
GLUCOSE
PYRUVIC ACID
GLUCONEOGENESIS
PYRUVIC ACID
GLUCOSE

GLYCOLYSIS IS ALL ABOUT BREAKING DOWN SACCHARIDES!
BREAK IT DOWN, Y'ALL!

BING!

OOH!
OOH!
I TOTALLY UNDERSTAND!
GLYCOLYSIS IS A PIECE OF CAKE!
...
...

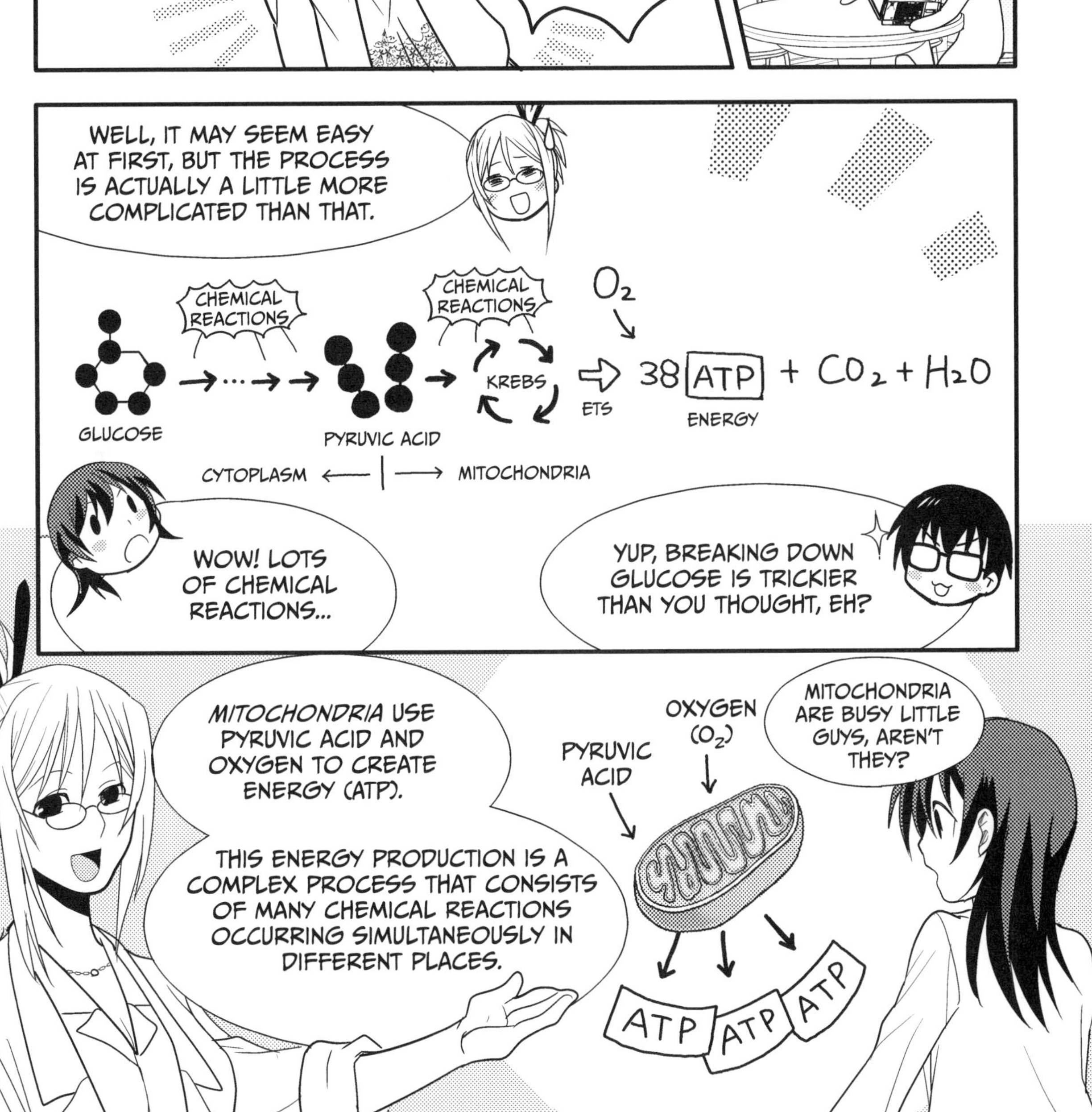
WELL, IT MAY SEEM EASY AT FIRST, BUT THE PROCESS IS ACTUALLY A LITTLE MORE COMPLICATED THAN THAT.
CHEMICAL REACTIONS
CHEMICAL REACTIONS
O_2
GLUCOSE
PYRUVIC ACID
KREBS
ETS
38 ATP + CO_2 + H_2O
ENERGY
CYTOPLASM
MITOCHONDRIA
WOW! LOTS OF CHEMICAL REACTIONS...
YUP, BREAKING DOWN GLUCOSE IS TRICKIER THAN YOU THOUGHT, EH?
MITOCHONDRIA USE PYRUVIC ACID AND OXYGEN TO CREATE ENERGY (ATP).
THIS ENERGY PRODUCTION IS A COMPLEX PROCESS THAT CONSISTS OF MANY CHEMICAL REACTIONS OCCURRING SIMULTANEOUSLY IN DIFFERENT PLACES.
PYRUVIC ACID
OXYGEN (O_2)
ATP
ATP
ATP
MITOCHONDRIA ARE BUSY LITTLE GUYS, AREN'T THEY?

BIOCHEMISTRY OF PHOTOSYNTHESIS
FINALLY, LET'S LOOK AT PHOTOSYNTHESIS IN PLANTS.
A COMPLEX CHEMICAL REACTION OCCURS WHEN LIGHT STRIKES *CHLOROPLASTS* IN A PLANT'S CELLS.
CARBON DIOXIDE IS USED AS A RAW MATERIAL BY THAT CHEMICAL REACTION TO CREATE SACCHARIDES SUCH AS GLUCOSE...
CO_2
LIGHT
CHEMICAL REACTION
GLUCOSE
H_2O
CHLOROPLAST
SO, KUMI, WHAT HAVE YOU NOTICED ABOUT ALL OF THESE CELLULAR PROCESSES? DO THEY HAVE ANYTHING IN COMMON?
HUH? UM...WELL...
生化学
秘
ノート
① PROTEIN SYNTHE
② METABOLISM
③ ENERGY PRODUCTION
OSYNTHESIS
THEY'RE ALL CHEMICAL REACTIONS?
YOU GOT IT!
ビシッ
KUMI GETS A GOLD STAR!

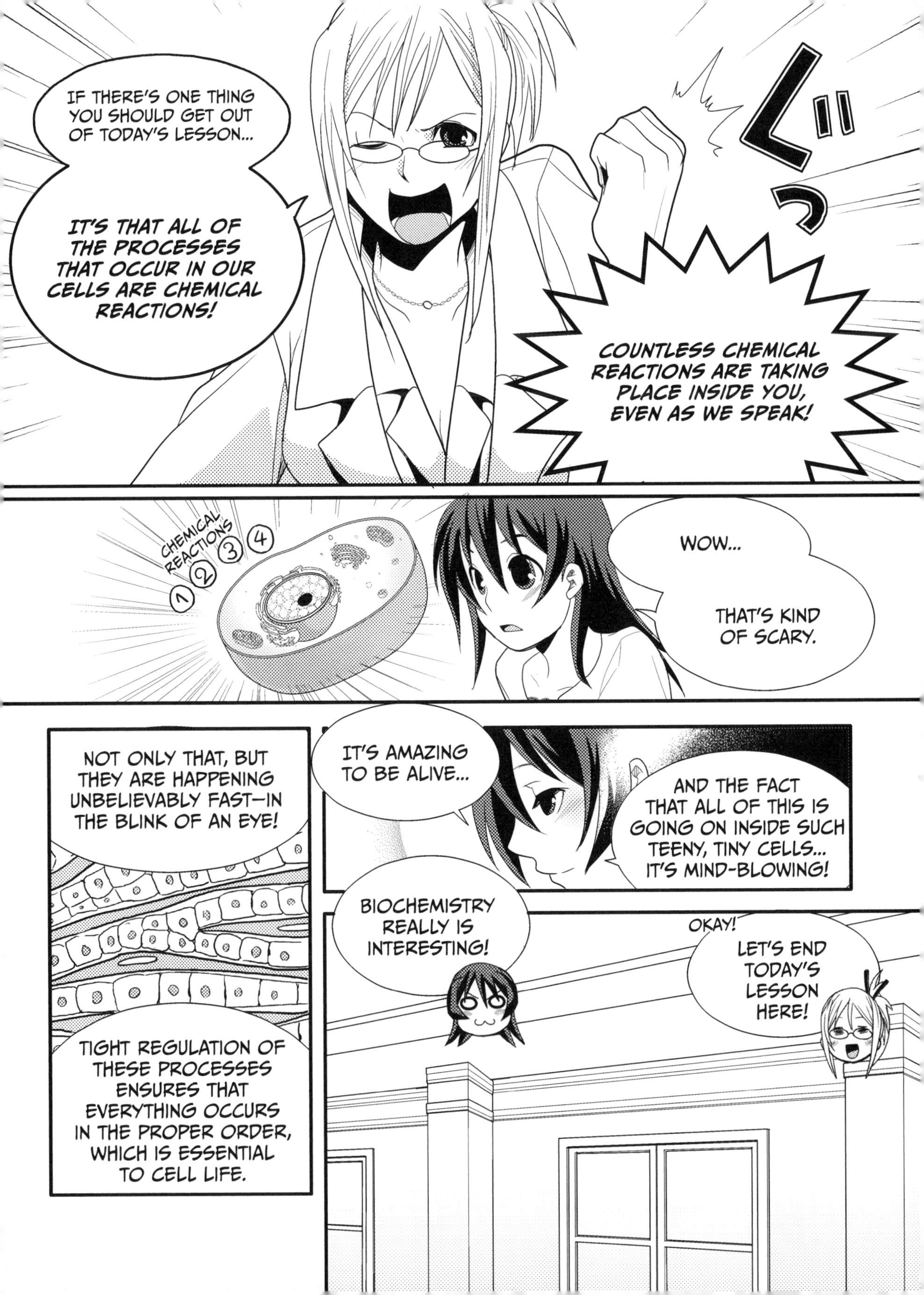
IF THERE'S ONE THING YOU SHOULD GET OUT OF TODAY'S LESSON...
IT'S THAT ALL OF THE PROCESSES THAT OCCUR IN OUR CELLS ARE CHEMICAL REACTIONS!
COUNTLESS CHEMICAL REACTIONS ARE TAKING PLACE INSIDE YOU, EVEN AS WE SPEAK!
CHEMICAL REACTIONS
1 2 3 4
WOW...
THAT'S KIND OF SCARY.
NOT ONLY THAT, BUT THEY ARE HAPPENING UNBELIEVABLY FAST—IN THE BLINK OF AN EYE!
TIGHT REGULATION OF THESE PROCESSES ENSURES THAT EVERYTHING OCCURS IN THE PROPER ORDER, WHICH IS ESSENTIAL TO CELL LIFE.
IT'S AMAZING TO BE ALIVE...
AND THE FACT THAT ALL OF THIS IS GOING ON INSIDE SUCH TEENY, TINY CELLS... IT'S MIND-BLOWING!
BIOCHEMISTRY REALLY IS INTERESTING!
OKAY!
LET'S END TODAY'S LESSON HERE!

CENSORED
ROBOCAT OUTPUT
WELL, IT'S BEEN A LONG DAY. YOU MUST BE WORN OUT.
DEFINITELY. BUT IT WAS REALLY VERY FASCINATING! THANK YOU SO MUCH!
SEE YOU LATER!
...NEMOTO?
SQUEEZE

ARE YOU SURE YOUR MIND IS ON BIOCHEMISTRY...
...AND NOT ON KUMI?

WHAT? NO WAY! I JUST WANT TO GET PEOPLE INTERESTED IN BIOCHEMISTRY, THAT'S ALL!
ひぃ〜〜
WHATEVER YOU SAY...
BUT I THINK YOU TWO HAVE SOME "CHEMISTRY" OF YOUR OWN.
THAT'S CRAZY! I'M JUST AN INNOCENT SCIENTIST! I WOULD NEVER-
アセ アセ
OKAY, NEMOTO, SAY NO MORE. I UNDERSTAND.
I HAVEN'T SEEN A BOY THIS HEAD-OVER-HEELS IN YEARS.
THERE'S ONLY ONE THING I CAN DO!
JUST CALL ME... THE PROFESSOR OF LOVE! ♪

4. Fundamental Biochemistry Knowledge

In this section, we'll explain some technical terms that you need to know to study biochemistry.

CARBON

First, we'll examine an extremely important chemical element in biochemistry—*carbon*.

Carbon is the element identified by chemical symbol C and possessing the atomic number 6 and an atomic weight of 12.0107. It's the primary component of all known life, which is why people sometimes refer to Earth's organisms as "carbon-based life." Carbon is the backbone of all organic compounds, and the bodies of living organisms are made almost entirely out of these compounds. Carbon is ideal as a backbone for complex organic molecules such as biopolymers, because it forms four stable bonds, which is an unusually high number for an element. Proteins, lipids, saccharides, nucleic acids, and vitamins are all built with carbon as a framework.

Although carbon is common on Earth—in the biosphere, lithosphere, atmosphere, and hydrosphere—there is a finite amount of it, so it's recycled and reused. Over time, a carbon atom passes through air, soil, rocks, and living creatures via biogeochemical cycles. The carbon in your body today may have once been inside a dinosaur!

CHEMICAL BONDS

When carbon combines with other elements, such as oxygen, hydrogen, or nitrogen, different chemical compounds are produced. Except for certain gases, like helium and argon, almost all chemical substances are composed of *molecules*, two or more atoms attached via a *chemical bond*. For example, a water molecule (H_2O) is created when two hydrogen atoms (H) and one oxygen atom (O) join together.

There are several different types of chemical bonds. Some examples include: *covalent bonds*, in which electrons are shared between a pair of atoms, *ionic bonds*, in which oppositely-charged atoms are attracted to one another, and *metallic bonds*, in which a pool of electrons swirl around numerous metal atoms.

The four stable bonds that carbon forms are all covalent bonds.

BIOPOLYMERS

Biopolymers are extremely important molecules to the study of biochemistry.

Biopolymer is a generic term for large, modular organic molecules. Modular means "assembled from repeating units," like the beads of a necklace. Proteins, lipids, nucleic acid, and polysaccharides are all biopolymers. Because they tend to be especially large molecules, biopolymers can form complex structures, which makes them very useful in advanced systems such as cells.

Biopolymers can form these complex chains because they're more than simple beads. Let's consider proteins, for example. Imagine a protein as a necklace made from a variety of different LEGO blocks that can all connect to one another. Since you can twist the necklace easily, it doesn't matter whether the blocks are close together or far apart, but the individual properties of each block result in some connecting better than others. If this necklace was a mile long, imagine the many strange and complex forms you could build. This isn't precisely how proteins function, but you get the idea.

ENZYMES

Since biochemistry explains life from a chemical point of view, it is vital to understand how chemical reactions work, and *enzymes* are essential to these reactions. Enzymes are proteins that act as catalysts—that is, they increase the rate of chemical reactions. An enzyme catalyzes nearly every chemical reaction that occurs in an organism.

In a chemical reaction catalyzed by an enzyme, the substance that the enzyme acts upon is called the *substrate*. The new substance that's formed during the reaction is called the *product*. The activity of an enzyme is affected by the environment inside the organism (temperature, pH, and other factors), the availability of the substrate, and, in some cases, the concentration of the product.

Although almost all enzymes are proteins, it has recently been discovered that a special type of ribonucleic acid (RNA) can act as a catalyst in certain chemical reactions. This is called an RNA enzyme, or a *ribozyme*.

OXIDATION-REDUCTION

Enzymes are broadly classified into six types, which will be introduced in detail in Chapter 4. *Oxidation-reduction* is one of the most important enzyme reactions, in which electrons are exchanged between two substances. If electrons are lost, the substance is *oxidized*, and if electrons are gained, the substance is *reduced*. Normally, when one substance is oxidized, another substance is reduced, so oxidation and reduction are said to occur simultaneously.

The movement of hydrogen ions (H^+, aka protons) often accompanies the exchange of electrons in an organism, and NADPH, NADH, and similar compounds (which we'll discuss in Chapter 2) work as *reducing agents* on other substances.

RESPIRATION

In Chapter 2, we will examine *respiration*. In the broadest sense, respiration is the process of obtaining energy by breaking down large compounds, but this only gives us a vague sense of the meaning.

More specifically, when respiration occurs, an organic substance (for example, the carbohydrates that make up spaghetti) is broken down into simple, inorganic components, like carbon dioxide (CO_2) and water (H_2O). Energy is produced when electrons are transferred between molecules (oxidation-reduction), along a sort of factory line, until they reach oxygen (O_2). This process is known as *internal respiration* or *cellular respiration*.

The oxygen we mentioned above is very important in respiration. It comes from the air that we breathe, and carbon dioxide is produced as a waste product of cellular respiration. When we use our lungs to inhale oxygen and exhale carbon dioxide, it's known as *external respiration*.

METABOLISM

The processes that alter an organism's chemical substances are called *metabolism*. Broadly speaking, metabolism can be divided into *substance metabolism* and *energy metabolism*. However, since these two types occur together during metabolism, the distinction isn't very clear. In this book, when we refer to metabolism, you may assume that we mean substance metabolism.

Substance metabolism This refers to the changes to substances that occur in an organism, including the chemical reactions that are catalyzed by enzymes. More specifically, a reaction that breaks down a complex substance into simpler substances is called *catabolism*, and, conversely, a reaction that synthesizes a more complex substance is called *anabolism*.

Energy metabolism This refers to the energy that's gained or lost through anabolic and catabolic processes within an organism, including reactions in which the energy created via respiration or photosynthesis is stored as ATP and other high-energy intermediates.

PHOTOSYNTHESIS AND RESPIRATION

1. Ecosystems and Cycles

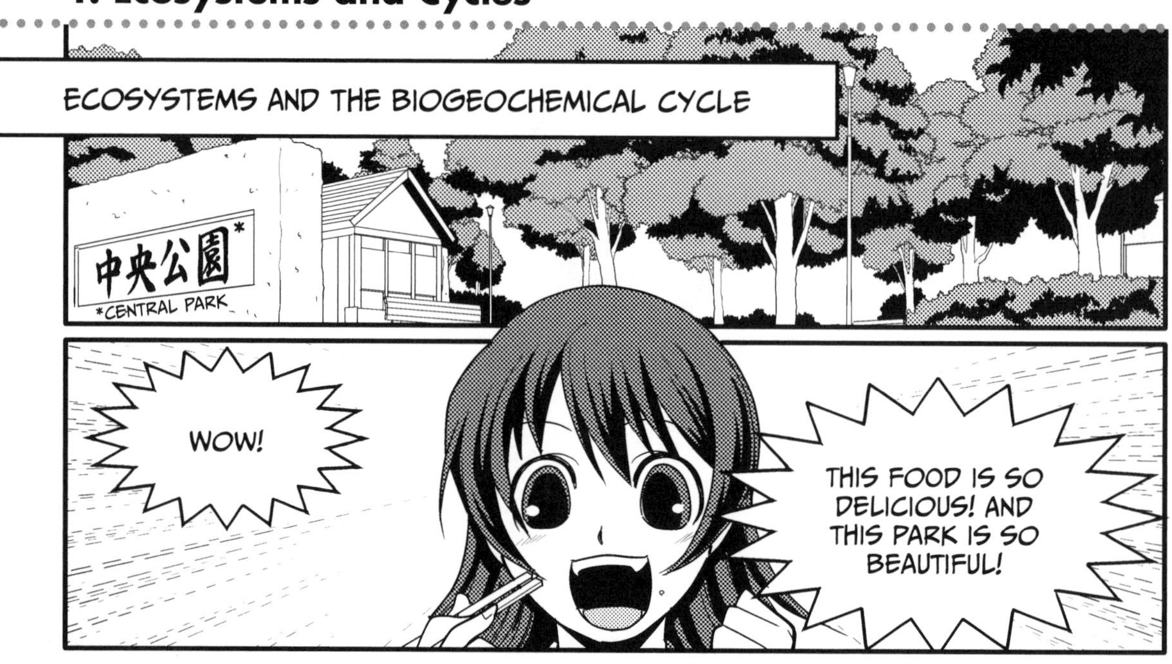

ABSOLUTELY! I'LL DO MY PART BY NOT USING AIR CONDITIONING!
INSTEAD, I'LL JUST EAT DELICIOUS ICE CREAM WHENEVER IT'S HOT OUTSIDE.
SO COOL, SO YUMMY!
THAT'S NOT EXACTLY WHAT I MEANT...
WELL, THE TERM "GLOBAL ENVIRONMENT" IS A LITTLE AMBIGUOUS...HOW AM I SUPPOSED TO KNOW EXACTLY WHAT YOU MEAN?
ARE YOU CONFUSED BECAUSE WE'RE TALKING ON SUCH A LARGE SCALE?
OF COURSE THAT'S HER PROBLEM! LET'S EXPLAIN TO KUMI WHAT MAKES UP THE EARTH'S ENVIRONMENT.
HARUMPH!
CLOSELY RELATED TO OUR ENVIRONMENT IS THE IDEA OF AN *ECOSYSTEM*.
THIS TERM COLLECTIVELY REFERS TO ALL THE PLANTS, ANIMALS, AND ORGANISMS THAT INHABIT A PARTICULAR AREA, ALONG WITH THE NONLIVING ELEMENTS OF THE SURROUNDING ENVIRONMENT...
IT INDICATES THAT THEY'RE ALL PART OF A SINGLE WHOLE, OR SYSTEM.

SINCE ALL LIFE ON EARTH IS INTERCONNECTED THROUGH NATURAL CYCLES...
...IT'S NOT AN EXAGGERATION TO CONSIDER THE GLOBAL ENVIRONMENT AS ITS OWN MASSIVE ECOSYSTEM.
WELL, I'LL BE...
BUT HOW CAN SUCH A VAST ECOSYSTEM BE MAINTAINED?
THE FOOD CHAIN IS IMPORTANT—THAT INDICATES WHICH ORGANISMS EAT WHICH OTHER ORGANISMS, BUT...
...HERE WE NEED TO CONSIDER HOW AN ECOSYSTEM COMES ABOUT CHEMICALLY.
WE ARE TALKING ABOUT BIOCHEMISTRY, AFTER ALL!

EVERYTHING IS LINKED THROUGH THE *BIOGEOCHEMICAL CYCLE*.
BIOGEOCHEMICAL CYCLE
IT'S...SOMETHING THAT CYCLES AROUND AND AROUND?
YOU GOT IT!
LET'S TAKE A CLOSER LOOK AT BIOGEOCHEMICAL CYCLES.

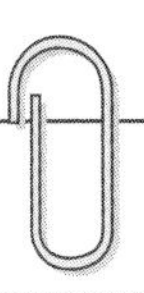

WHAT IS THE BIOGEOCHEMICAL CYCLE?

The elements of the global ecosystem, including the food chain, respiration, and photosynthesis, are all part of a worldwide cycle known as the biogeochemical cycle.

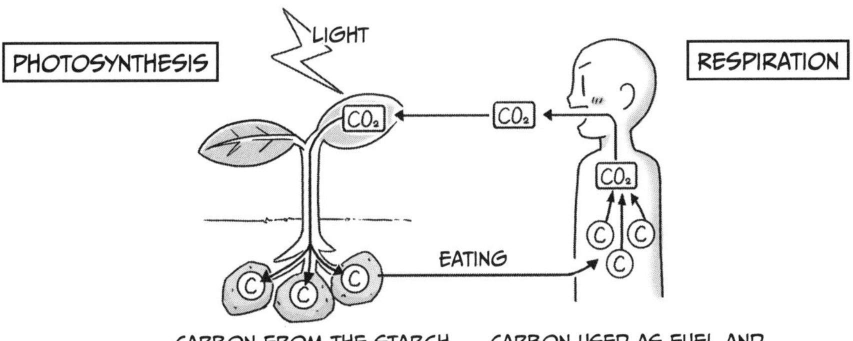

When one organism, like a bear, eats another organism, such as a fish, the material that made up the fish becomes part of the bear.

A very important substance in this transfer is **carbon (C)**.

Look at the diagram above. If Kumi eats a potato, the carbon from that potato enters her body.

And when I breathe, the carbon that was inside my body becomes carbon dioxide (CO_2) and is expelled from my body.

That's right! Then the carbon that left your body is captured by plants through photosynthesis and is transformed into the carbon that makes up starch (which is a type of saccharide).

Then that potato is eaten by Kumi (or a cow or some other hungry creature), and the carbon is returned once more to the body of an animal.

Since Kumi eats beef, the carbon can also move from the cow to Kumi.

Mmm...I love a tasty cheeseburger!

So carbon really moves around a lot, doesn't it? Is this movement the biogeochemical cycle?

Yep! The entire Earth is carrying out this cycle on a gigantic scale.

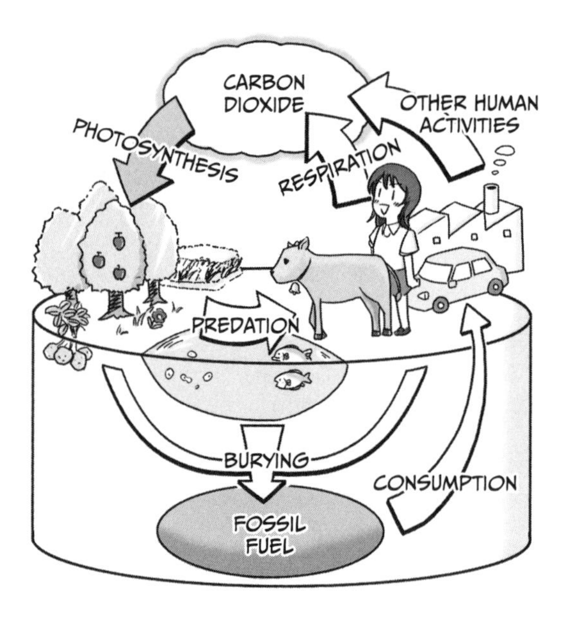

Wow! So the rice, potatoes, and apples that I eat were originally related to carbon dixiode that somebody else exhaled.

And it's not just carbon. Hydrogen (H), oxygen (O), nitrogen (N), and sulfur (S) cycle around the Earth as well, going from organism to organism, getting emitted into the atmosphere, dissolved into the ocean, or buried deep underground.

When this cycle works smoothly, the ecosystem and the global environment are healthy.

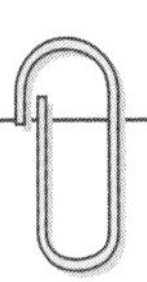

CARBON CYCLE

Next I'll explain the carbon cycle in a little more detail.

Umm, carbon...? I know we've talked a little about it, but I still don't understand what it is.

No problem! Nemoto, can you review some basic facts about carbon for Kumi?

Sure. Carbon is one of the most important chemical elements to living creatures like us.

Carbon is at the center of amino acids, which are the building blocks of proteins. It's also the element that creates the framework of saccharides and lipids, and it's a vital part of DNA.

H_2N—C—COOH (H above C, R below C)

IT'S HERE.

IT'S ALSO OFTEN HERE.

AMINO ACID

CH_2OH

GLUCOSE

H—C—C—C—C—C—C … C—C—COOH

YOU'LL EVEN FIND SOME HERE.

FATTY ACID

You see? C is essential to all of them.

You're right. It looks like we'd be in real trouble without any C. It really is important!

When carbon leaves an organism's body, it can bond with two oxygen atoms to become carbon dioxide (CO_2) or bond with four hydrogen atoms to become methane (CH_4).

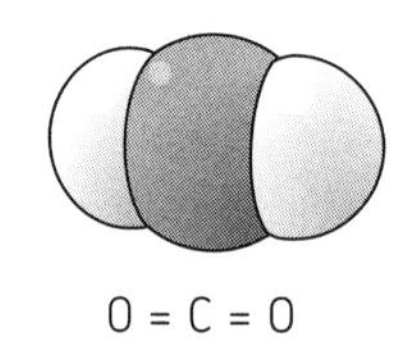

Carbon dioxide

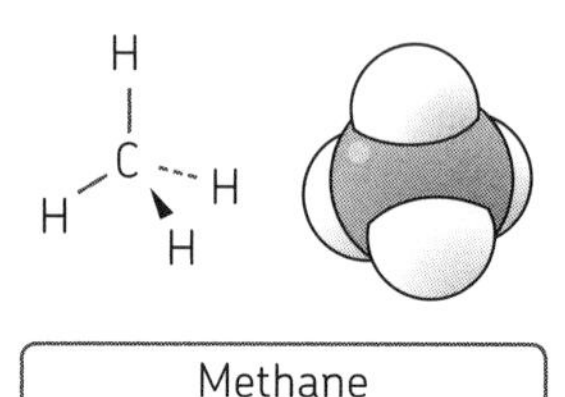

Methane

Also, it can accumulate deep underground and, over long periods of time, it will become crude oil, coal, and, in some cases, even diamonds.

Oil! Diamonds! Wow, carbon is really valuable stuff!

That's true, but remember: It's not all about excitement and riches. The way carbon circulates is extremely important. Disrupting this balance could make carbon dioxide concentration on Earth steadily increase...

Which would be a serious problem, wouldn't it?

Oh man, that sounds like it would be a total bummer...

Well, think of it this way: If you can understand the Earth as a single circulating system and your own body as a system as well, then you'll realize the importance of a healthy balance and the dangers of disrupting that balance. This knowledge can lead to insights into the pursuit of health and beauty!

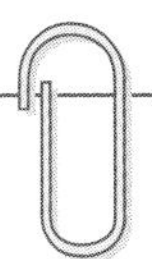

The pursuit of health and beauty? Now you've got my attention!

(Wow, Professor Kurosaka is really good at motivating Kumi...)

Let's look at this figure again.

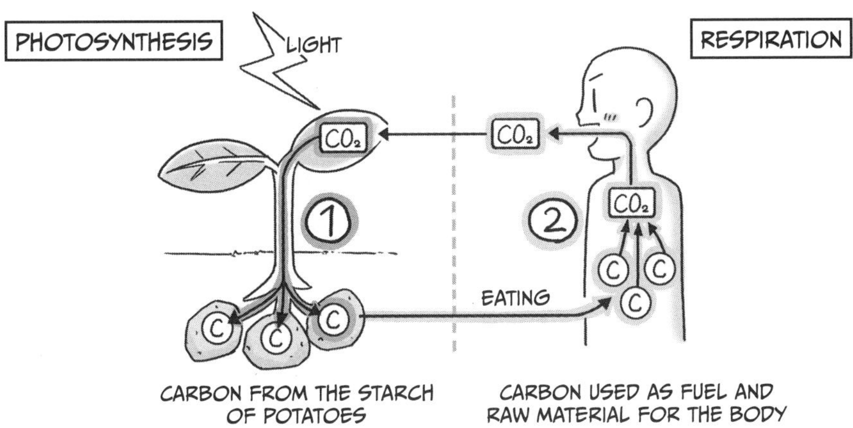

Look at ① on the left. This is the **flow of carbon in which plants pull carbon dioxide from the atmosphere during photosynthesis and use it as a raw material for creating saccharides.**

Now look at ② on the right. This is the **flow of carbon in which those saccharides are used by a living creature, and through this process, they become carbon dioxide once more and are returned to the atmosphere via respiration.**

Let's talk more about these two flows. Now that we're finished with lunch, we can really sink our teeth into the subject!

Super cool!

2. Let's Talk Photosynthesis

THE IMPORTANCE OF PLANTS

At the base of the ecosystem, plants, algae, and some bacteria supply food to all living creatures. The majority of them use a process called *photosynthesis*, in which saccharides and oxygen (O_2) are created from carbon dioxide by splitting atoms of water, using energy from the Sun. Saccharides are also known as *carbohydrates* and are vital to living creatures like us.

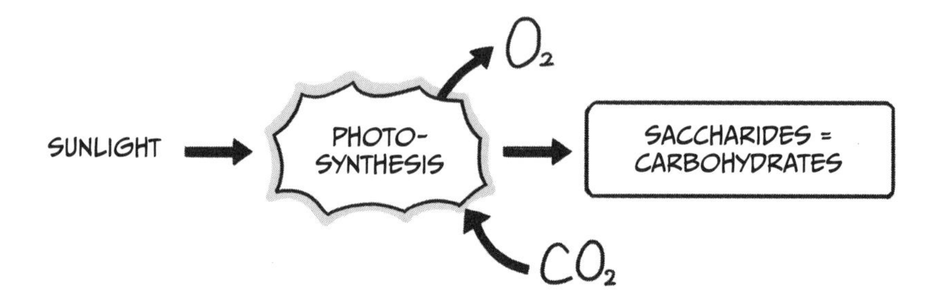

That's why plants are called *producers*. In contrast, we animals are called *consumers*.

Photosynthesis is important not just because it creates saccharides. It also maintains a steady, balanced concentration of carbon dioxide in the atmosphere, and it produces oxygen, which living animals require to survive.

As you can imagine, deforestation by humans greatly reduces the number of "producers," upsetting the delicate balance of the biogeochemical cycle and potentially threatening "consumers" (like us!).

Now, let's take a closer look at the way in which plants use sunlight to create saccharides.

CHLOROPLAST STRUCTURE

This image (courtesy of RoboCat) shows green sacks, called *chloroplasts*, within a plant cell.

Inside a chloroplast, structures shaped like very thin pouches are stacked in multiple layers to form peculiar structures. Each of these flat pouches is called a *thylakoid*, and a stack of multiple thylakoids is called a *granum*.

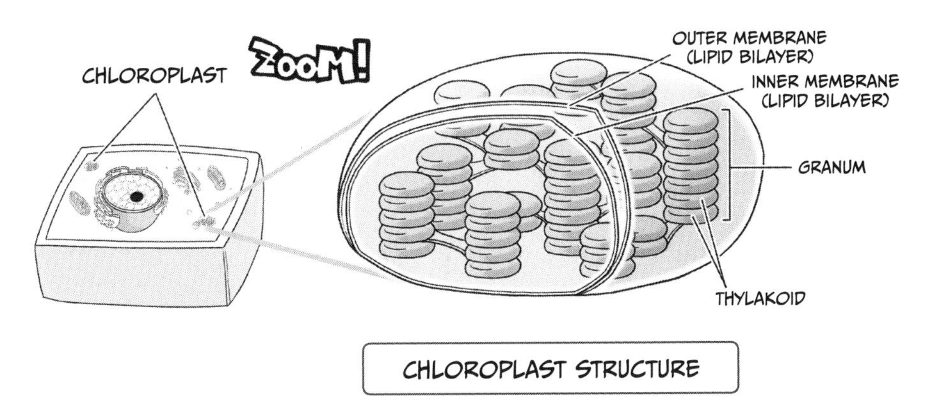

CHLOROPLAST STRUCTURE

The thylakoid membrane is a bilayer composed primarily of phospholipids, just like in cell membranes.

Now let's look at the surface of the thylakoid membrane.

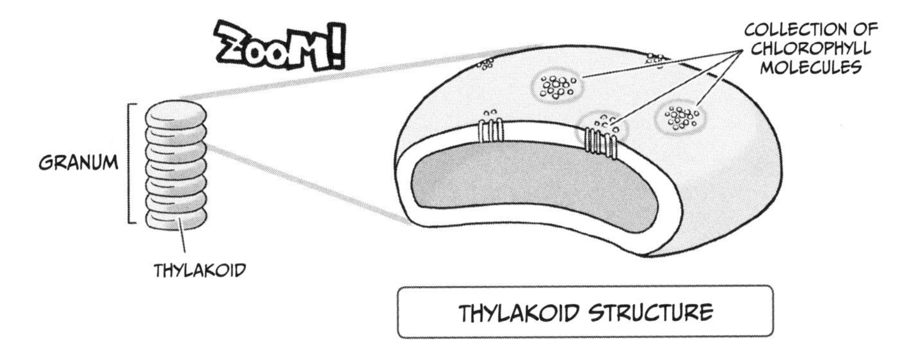

THYLAKOID STRUCTURE

Do you see the groups of tiny grains that appear to be embedded in the surface of the thylakoid? Each of these is a collection of molecules called *chlorophyll*, as well as various proteins that aid in photosynthesis.

Chlorophyll molecules absorb sunlight. However, they don't absorb the entire spectrum; they reflect the green portion of sunlight back outside, which is why plants appear green to us.

PHOTOSYNTHESIS—THE PHOTOPHOSPHORYLATION REACTION
WELL, SINCE WE WERE ABLE TO FIND SUCH A NICE GAZEBO...
WHY DON'T WE CONTINUE OUR LESSONS OUT HERE IN THE PARK?
YAY!!
UM, ARE YOU SURE THE LAB WOULDN'T BE BETTER?
BUT WE'RE GOING TO DISCUSS PHOTOSYNTHESIS!
DON'T YOU THINK IT'S MORE FITTING TO STUDY UNDER A BLUE SKY THAN TO SHUT OURSELVES OFF IN THE LABORATORY?
I GUESS...
OH, I ALREADY LEARNED ABOUT PHOTOSYNTHESIS! HUMANS CAN'T DO IT—ONLY PLANTS!
THEY USE SUNLIGHT AND CARBON DIOXIDE TO CREATE OXYGEN AND SACCHARIDES, RIGHT?
PHOTOSYNTHESIS
SACCHARIDE = CARBOHYDRATE
O_2
YES, THAT'S CORRECT!
BUT DO YOU KNOW *HOW* PLANTS PRODUCE OXYGEN AND SACCHARIDES FROM SUNLIGHT? DO YOU KNOW HOW THE PROCESS OF PHOTOSYNTHESIS WORKS?
URK

THE TWO MOST IMPORTANT REACTIONS IN PHOTOSYNTHESIS ARE *PHOTOPHOSPHORYLATION* AND *CARBON DIOXIDE FIXATION*.
PHOTOPHOSPHORYLATION IS THE CREATION OF ENERGY USING SUNLIGHT.
CARBON DIOXIDE FIXATION USES THAT ENERGY TO SYNTHESIZE SACCHARIDES.
キラ キラ
SUNLIGHT REQUIRED!
PHOTOPHOS-PHORYLATION
CARBON DIOXIDE FIXATION
H_2O
O_2
CO_2
SACCHA-RIDE
FIRST, WE'LL LEARN ABOUT PHOTO-PHOSPHORYLATION.
LET'S INVESTIGATE WHERE PHOTO-PHOSPHORYLATION IS PERFORMED INSIDE A CHLOROPLAST.
ドキ ドキ
WE CAN USE ROBOCAT TO DIVE IN AND EXPLORE THE INSIDE OF A PLANT!
THIS IS AN IMAGE LOOKING DOWN AT A THYLAKOID INSIDE A CHLOROPLAST.
SEE THE COLLECTIONS OF BUMPS?
EACH OF THESE IS AN AGGREGATION OF MOLECULES CALLED CHLOROPHYLL AND VARIOUS PROTEINS.
PING!
DOZENS OF CHLOROPHYLL MOLECULES ARE CONTAINED IN EACH GROUP, AND EACH OF THESE MOLECULES ABSORBS SUNLIGHT LIKE A LITTLE SATELLITE DISH.
ERRR...

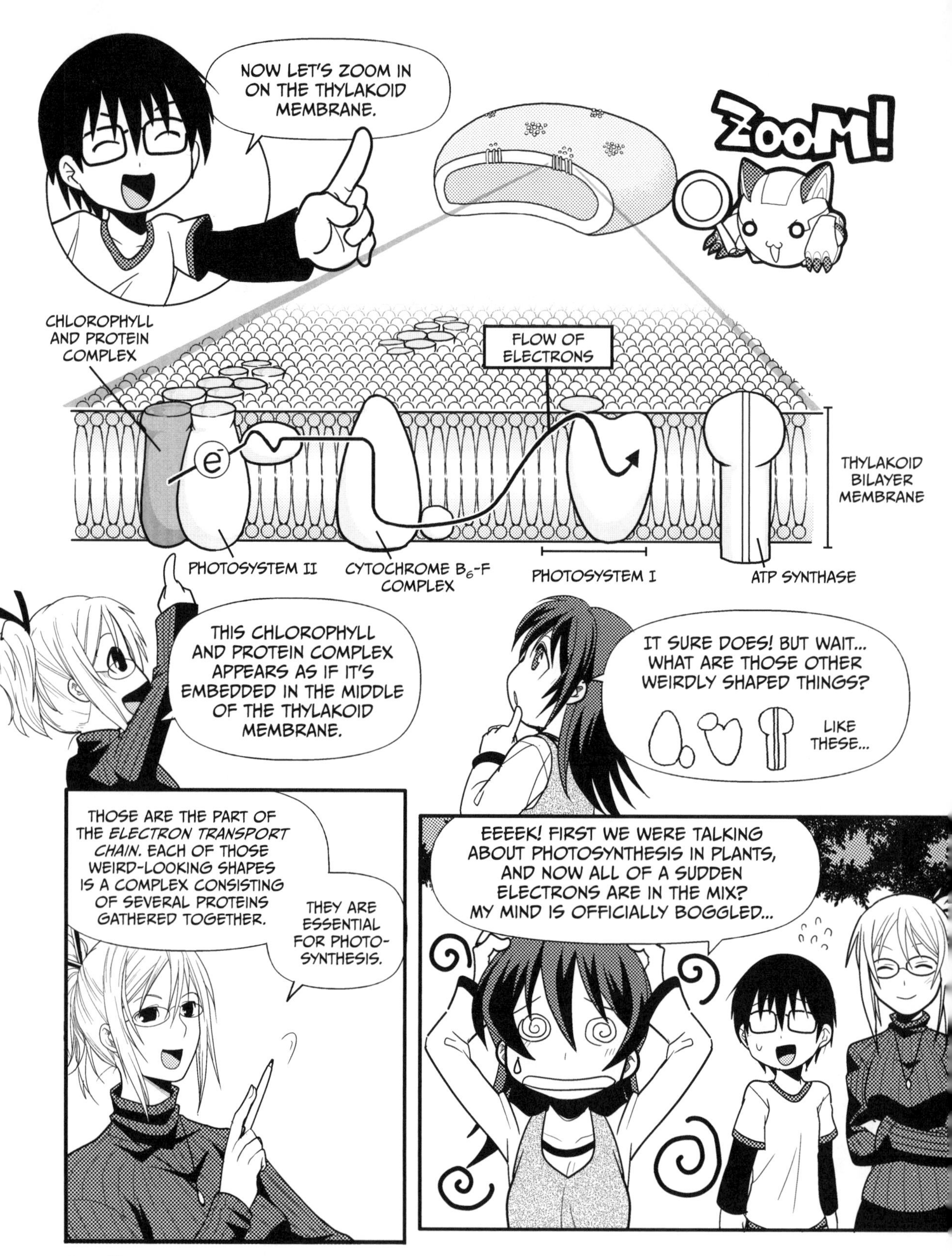
NOW LET'S ZOOM IN ON THE THYLAKOID MEMBRANE.
ZOOM!
CHLOROPHYLL AND PROTEIN COMPLEX
FLOW OF ELECTRONS
e⁻
THYLAKOID BILAYER MEMBRANE
PHOTOSYSTEM II
CYTOCHROME B6-F COMPLEX
PHOTOSYSTEM I
ATP SYNTHASE
THIS CHLOROPHYLL AND PROTEIN COMPLEX APPEARS AS IF IT'S EMBEDDED IN THE MIDDLE OF THE THYLAKOID MEMBRANE.
IT SURE DOES! BUT WAIT... WHAT ARE THOSE OTHER WEIRDLY SHAPED THINGS?
LIKE THESE...
THOSE ARE THE PART OF THE *ELECTRON TRANSPORT CHAIN*. EACH OF THOSE WEIRD-LOOKING SHAPES IS A COMPLEX CONSISTING OF SEVERAL PROTEINS GATHERED TOGETHER.
THEY ARE ESSENTIAL FOR PHOTO-SYNTHESIS.
EEEEK! FIRST WE WERE TALKING ABOUT PHOTOSYNTHESIS IN PLANTS, AND NOW ALL OF A SUDDEN ELECTRONS ARE IN THE MIX? MY MIND IS OFFICIALLY BOGGLED...

* ACTUALLY, ELECTRONS FLY OFF ONLY FROM THE CENTRAL *REACTION CENTER CHLOROPHYLL*. THE OTHER CHLOROPHYLL MOLECULES ACT AS AMPLIFIERS TO FEED ENERGY INTO THE REACTION CENTER, WHICH ENABLES THE ELECTRONS TO DETACH.

THE PROTEIN COMPLEXES DO MORE THAN TRANSPORT ELECTRONS IN A PREDETERMINED ORDER.
A MOLECULE CALLED *NADPH* IS SYNTHESIZED ALONG THE WAY, AND *ATP* IS CREATED AT THE END OF THE CHAIN.
ATP
YOU MENTIONED ATP BEFORE, BUT I THINK THIS IS THE FIRST TIME I'VE HEARD OF NADPH...
ATP
NADPH
NADPH CONSISTS OF AN ELECTRON AND A PROTON (HYDROGEN ION) ADHERED TO A MOLECULE CALLED A *HYDROGEN ACCEPTOR*. IT IS CREATED BY NADPH REDUCTASE.
PROTON
e
NADPH
TOSS 'EM IN HERE!
NADPH CAN BE THOUGHT OF AS A TEMPORARY STOREHOUSE FOR AN ELECTRON AND A PROTON, WHICH ARE REQUIRED BY THE CARBON DIOXIDE FIXATION REACTION WE TALKED ABOUT EARLIER.*
* IN OTHER WORDS, $NADP^+$ IS "REDUCED," AND NADPH IS CREATED. (SEE PAGE 37 FOR AN EXPLANATION OF REDUCTION.)
THE ENERGY THAT'S GENERATED IS JUST A "FLOW OF ELECTRONS," AND, IN THAT SENSE, IT'S JUST LIKE THE ELECTRICITY YOU USE FOR YOUR HOUSEHOLD APPLIANCES.
BREAKFAST TIME!
THE ELECTRON TRANSPORT CHAIN ALSO CREATES A "FLOW OF ELECTRONS." SUBSTANCES LIKE ATP ARE SYNTHESIZED BY THIS ELECTRON FLOW.
NOW LET'S LOOK AT THE WAY ATP IS CREATED BY PHOTO-PHOSPHORYLATION!
FWIP

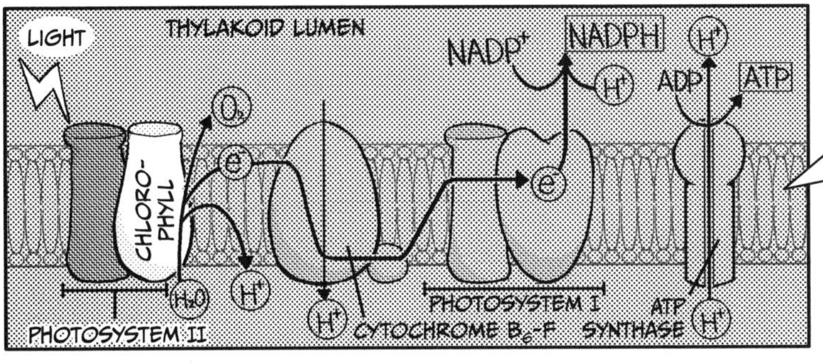

STEP 1 (PHOTOSYSTEM II)

SUNLIGHT STRIKES THE CHLOROPHYLL.*

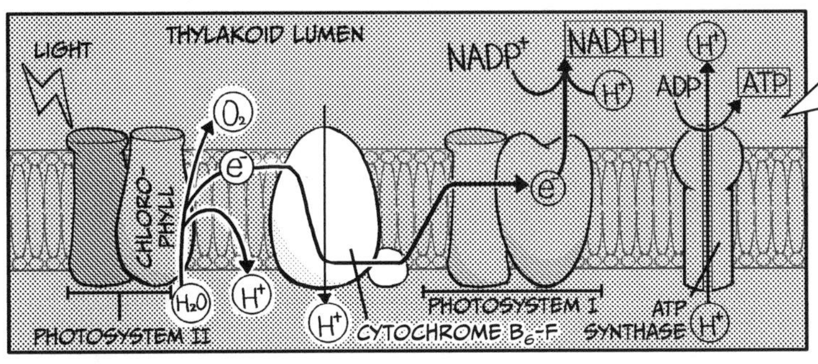

STEP 2

THE LIGHT ENERGY CAUSES AN EXCITED STATE IN THE CHLOROPHYLL SO AN ELECTRON e^- IS EMITTED AND PASSED ALONG. AT THE SAME TIME, A PROTON H^+ ACCUMULATES IN THE THYLAKOID LUMEN.

e^- STANDS FOR THIS ELECTRON.

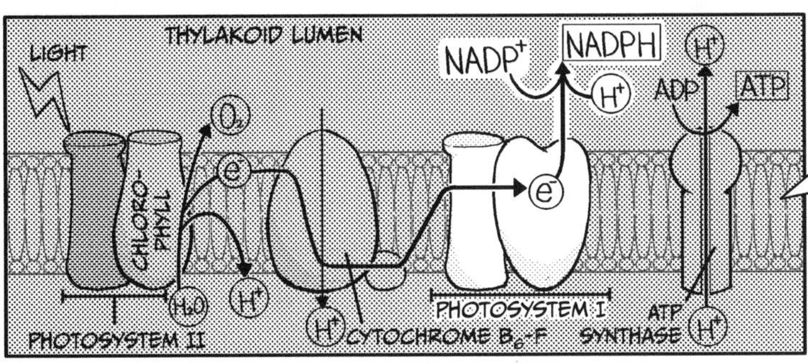

STEP 3 (PHOTOSYSTEM I)

THE ELECTRON (e^-) AND A PROTON ARE CAPTURED BY $NAPD^+$, AND NADPH IS PRODUCED.

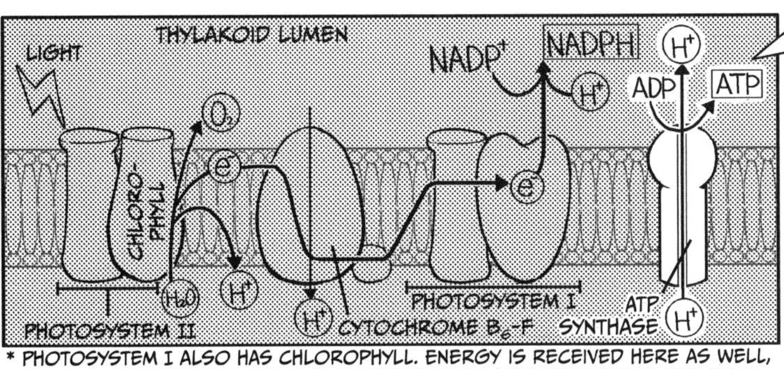

STEP 4

WHEN THE PROTONS (H^+) THAT WERE COLLECTED IN THE THYLAKOID LUMEN ARE ABOUT TO LEAVE THE THYLAKOID ACCORDING TO THE CONCENTRATION GRADIENT,** THEY PASS INTO ATP SYNTHASE. AT THIS TIME, ATP IS SYNTHESIZED FROM ADP.

** A FORCE WHICH CAUSES A SUBSTANCE TO NATURALLY FLOW FROM A HIGH CONCENTRATION TO A LOW CONCENTRATION.

* PHOTOSYSTEM I ALSO HAS CHLOROPHYLL. ENERGY IS RECEIVED HERE AS WELL, AND ELECTRONS THAT WERE TRANSPORTED FROM PHOTOSYSTEM II ARE ONCE AGAIN EXCITED.

OH, I GET IT! WHEN I SEE IT LAID OUT STEP-BY-STEP, IT MAKES WAY MORE SENSE!

BECAUSE OF THIS "ELECTRON FLOW," NADPH AND ATP ARE PRODUCED!

THE IMPORTANT THING HERE IS THE REACTION THAT CREATES THIS ATP, CALLED *PHOSPHORYLATION*.
PHOSPHORYLATION? WHAT KIND OF REACTION IS THAT?
WELL, LOOK...HERE'S THE STRUCTURE OF ADP (ADENOSINE DIPHOSPHATE).
O
C
C
TWO PHOSPHATE GROUPS
ADP
ADP, WHICH HAS TWO PHOSPHATE GROUPS, BECOMES ATP (ADENOSINE TRIPHOSPHATE) IF ONE ADDITIONAL PHOSPHATE GROUP IS ATTACHED.
O
C
C
C
ONE MORE PHOSPHATE GROUP
ATP
THE REACTION THAT ATTACHES THIS NEW PHOSPHATE GROUP IS PHOSPHORYLATION!
P
OH, SO THAT MUST BE WHERE THE NAME "PHOTOPHOSPHORYLATION" COMES FROM!
REMEMBER, ATP IS ALSO THE "COMMON CURRENCY" USED IN REACTIONS OTHER THAN PHOTOSYNTHESIS. SIMPLY PUT, IT'S IMPORTANT CHEMICAL ENERGY!
LET ME SUMMARIZE PHOTOPHOSPHORYLATION FOR YOU!
IT'S A REACTION THAT USES
LIGHT
TO CREATE ATP FROM ADP BY PHOSPHORYLATION
SO ATP IS SYNTHESIZED BY THE FLOW OF ELECTRONS TRIGGERED BY SUNLIGHT...
I UNDERSTAND! LIGHT ENERGY TURNS INTO CHEMICAL ENERGY!

PHOTOSYNTHESIS—CARBON DIOXIDE FIXATION
PHOTOPHOSPHORYLATION
CARBON FIXATION
THAT'S ALL FOR PHOTO-PHOSPHORYLATION.
JUST REMEMBER: SUNLIGHT IS ALWAYS REQUIRED!
NEXT, WE'LL STUDY *CARBON FIXATION.*
O-OKAY!
THE FINAL STEPS OF PHOTO-PHOSPHORYLATION IN WHICH ATP AND NADPH ARE SYNTHESIZED WILL OCCUR EVEN IF LIGHT IS NO LONGER PRESENT. THIS IS KNOWN AS A LIGHT-INDEPENDENT REACTION.*
ATP
NADPH
THEN DOES CARBON DIOXIDE FIXATION HAPPEN EVEN AT NIGHT OR ON A CLOUDY DAY?
ACTUALLY, NO. IT JUST MEANS THAT LIGHT ISN'T NEEDED AT THAT PARTICULAR STATION. THE REACTION ITSELF OCCURS IN THE DAYTIME BECAUSE NADPH IS REQUIRED AND IS NOT MANUFACTURED AT NIGHT.
* THIS LIGHT-INDEPENDENT REACTION IS ALSO KNOWN AS THE *CALVIN CYCLE* OR THE *REDUCTIVE PENTOSE PHOSPHATE CYCLE*.
CARBON DIOXIDE FIXATION OCCURS IN THE *STROMA*, WHICH IS THE CENTRAL PART OF THE CHLOROPLAST, RATHER THAN IN THE THYLAKOID MEMBRANE.
SEE? RIGHT HERE.
GOTCHA.
STROMA

CARBON FIXATION IS A REACTION THAT USES CHEMICAL ENERGY, STORED AS ATP, TO CREATE SACCHARIDES, USING THE CARBON DIOXIDE IN THE AIR AS A RAW MATERIAL.
ATP
ADP
NADPH
NADP+
CO2
SACCHARIDES
SketchBook
THAT'S THE CARBON DIOXIDE THAT'S PRODUCED BY HUMANS AND OTHER ANIMALS, ISN'T IT?
AND THE REACTION ALSO NEEDS THE CHEMICAL ENERGY THAT WAS CREATED BY PHOTOPHOSPHORYLATION.
ATP
NADPH
CARBON FIXATION
CO2

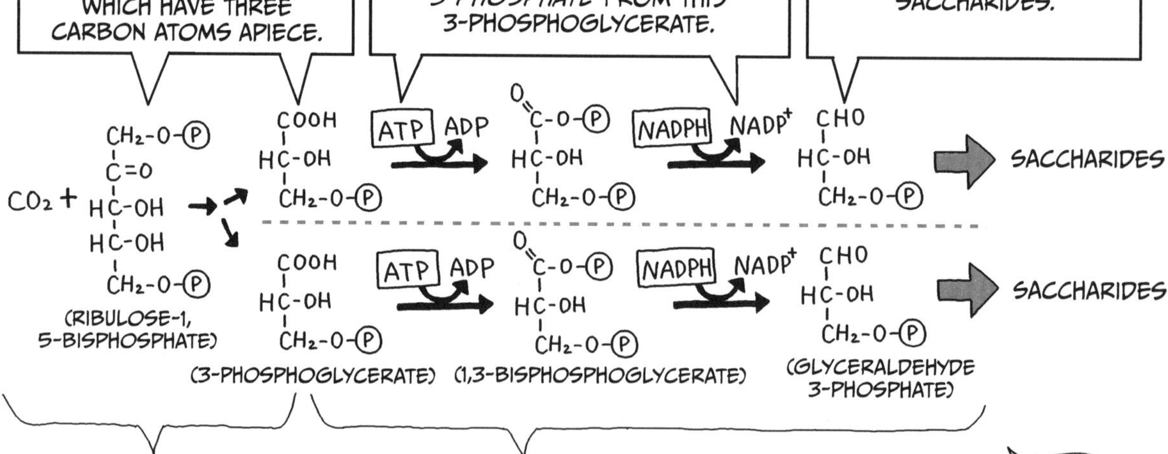
FIRST, CO2 BONDS WITH A SUBSTANCE CALLED *RIBULOSE-1, 5-BISPHOSPHATE* TO FORM TWO MOLECULES OF *3-PHOSPHOGLYCERATE*, WHICH HAVE THREE CARBON ATOMS APIECE.
THE CHEMICAL ENERGY OF ATP AND THE REDUCING POWER OF NADPH ARE USED TO CREATE TWO MOLECULES OF *GLYCERALDEHYDE 3-PHOSPHATE* FROM THIS 3-PHOSPHOGLYCERATE.
AND THIS GLYCERALDEHYDE 3-PHOSPHATE IS USED TO PRODUCE GLUCOSE, FRUCTOSE, AND OTHER SACCHARIDES.
CO2 +
CH2-O-P
C=O
HC-OH
HC-OH
CH2-O-P
(RIBULOSE-1, 5-BISPHOSPHATE)
COOH
HC-OH
CH2-O-P
(3-PHOSPHOGLYCERATE)
ATP
ADP
O
C-O-P
HC-OH
CH2-O-P
(1,3-BISPHOSPHOGLYCERATE)
NADPH
NADP+
CHO
HC-OH
CH2-O-P
(GLYCERALDEHYDE 3-PHOSPHATE)
SACCHARIDES
SACCHARIDES
BONDING OF CARBON DIOXIDE
USE OF CHEMICAL ENERGY

SO CARBON DIOXIDE IS BOUND FIRST, AND THEN CHEMICAL ENERGY IS USED.

* MORE PRECISELY, THE SACCHARIDE CREATED BY PHOTOSYNTHESIS IS GLUCOSE 1-PHOSPHATE.

** PLANTS SYNTHESIZE SUCROSE FROM GLUCOSE TO MAKE THEM MORE APPEALING FOR ANIMALS TO EAT.

3. Respiration

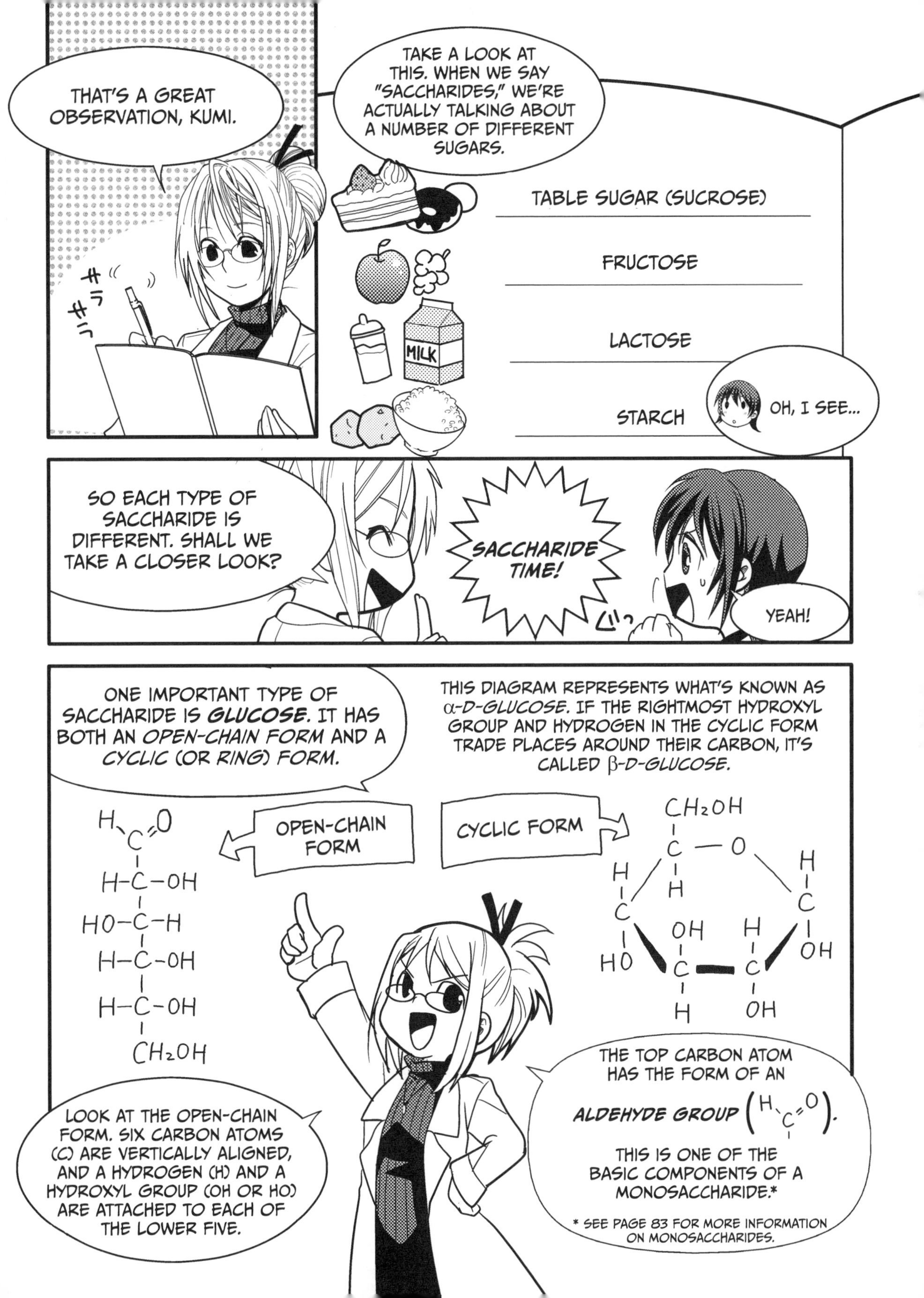
THAT'S A GREAT OBSERVATION, KUMI.
TAKE A LOOK AT THIS. WHEN WE SAY "SACCHARIDES," WE'RE ACTUALLY TALKING ABOUT A NUMBER OF DIFFERENT SUGARS.
TABLE SUGAR (SUCROSE)
FRUCTOSE
LACTOSE
MILK
STARCH
OH, I SEE...
SO EACH TYPE OF SACCHARIDE IS DIFFERENT. SHALL WE TAKE A CLOSER LOOK?
SACCHARIDE TIME!
YEAH!
ONE IMPORTANT TYPE OF SACCHARIDE IS GLUCOSE. IT HAS BOTH AN OPEN-CHAIN FORM AND A CYCLIC (OR RING) FORM.
THIS DIAGRAM REPRESENTS WHAT'S KNOWN AS α-D-GLUCOSE. IF THE RIGHTMOST HYDROXYL GROUP AND HYDROGEN IN THE CYCLIC FORM TRADE PLACES AROUND THEIR CARBON, IT'S CALLED β-D-GLUCOSE.
OPEN-CHAIN FORM
CYCLIC FORM
H C O
H-C-OH
HO-C-H
H-C-OH
H-C-OH
CH2OH
CH2OH
C O
H H H
C C
HO OH H OH
C C
H OH
LOOK AT THE OPEN-CHAIN FORM. SIX CARBON ATOMS (C) ARE VERTICALLY ALIGNED, AND A HYDROGEN (H) AND A HYDROXYL GROUP (OH OR HO) ARE ATTACHED TO EACH OF THE LOWER FIVE.
THE TOP CARBON ATOM HAS THE FORM OF AN ALDEHYDE GROUP (H C O).
THIS IS ONE OF THE BASIC COMPONENTS OF A MONOSACCHARIDE.*
* SEE PAGE 83 FOR MORE INFORMATION ON MONOSACCHARIDES.

SUGAR DISSOLVES EASILY, RIGHT? IT CAN TRAVEL THROUGHOUT THE BODY VIA THE BLOODSTREAM BECAUSE HYDROXYL GROUPS BLEND WELL WITH WATER.
OH
SLOSH
H2O
CLEARLY, THE PRESENCE OF OH IS AN IMPORTANT COMPONENT OF SUGAR, BUT WE ALSO NEED TO KNOW WHERE ON THE SACCHARIDE THE OH IS LOCATED.
C-OH
REALLY? WHY?
GLUCOSE
GALACTOSE
WELL, IF THE OH ATTACHED TO THE FOURTH CARBON IS ON THE LEFT SIDE...
...IT WILL NO LONGER BE GLUCOSE BUT RATHER A DIFFERENT TYPE OF SACCHARIDE CALLED GALACTOSE.
IT BECOMES SOMETHING DIFFERENT DEPENDING ON WHETHER IT'S ATTACHED ON THE RIGHT OR LEFT...THAT'S SO WEIRD.
ほお～
MONOSACCHARIDES, WHEN IN AN ACTUAL BODY, TAKE THE CYCLIC FORM MORE OFTEN THAN THE OPEN-CHAIN FORM.
GLUCOSE IS ALSO SHAPED LIKE:
MORE OFTEN THAN:
HOW'D SHE SAY THAT?
SUGAR

SACCHARIDES AND THE "-OSE" SUFFIX

You've probably noticed that most of the saccharides we've discussed so far, like glucose and galactose, end with the suffix "-ose." There are standardized rules for naming saccharides, so these names usually end with "-ose."

For instance, glucose is the saccharide that's the basis of energy production, and it's the sugar referred to when we talk about *blood sugar*. Common table sugar is technically called *sucrose*. Milk contains a saccharide called *milk sugar*, which is known as *lactose*, and the sugar contained in fruit is called *fructose*. It's important to remember that there are several kinds of saccharides in the natural world, and the structures of sucrose, lactose, glucose, galactose, and fructose actually differ somewhat. Also, the starch found in rice, potatoes, and other starchy foods is made from *amylose* and *amylopectin*.

In Chapter 3, we'll examine the structures of these saccharides in detail.

WHY DO MONOSACCHARIDES TAKE A CYCLIC STRUCTURE?

Why do monosaccharides take a cyclic structure more often than an open-chain structure? The secret is in an OH that's bonded to a carbon in the molecule.

Alcohol is a good example of this: All types of alcohol are represented in the form R-OH (where R is the variable group). Alcohol can bind with an aldehyde group or a ketone group to create a substance called *hemiacetal*. Since an OH of a monosaccharide also has this property, the monosaccharide ends up reacting with an aldehyde group or a ketone group within the molecule, and a cyclic structure is formed as the result.

ALCOHOL + ALDEHYDE ⇌ HEMIACETAL

THE OH IN THE FIFTH POSITION OF THE MONOSACCHARIDE REACTS WITH THE ALDEHYDE...

...TO FORM A CYCLICAL STRUCTURE.

WHY DO WE NEED TO BREATHE?
AS YOU KNOW, HUMAN BEINGS ARE CONSTANTLY BREATHING...
...BUT DO YOU KNOW WHY?
YES! IT'S BECAUSE WE MUST TAKE IN THE OXYGEN CONTAINED IN THE AIR AND EXPEL CARBON DIOXIDE.
O_2
CO_2
WHA?!

THAT'S RIGHT. BUT, MORE SPECIFICALLY, OUR CELLS REQUIRE OXYGEN TO CREATE THE ENERGY THEY NEED TO KEEP US ALIVE.
ATP
"GOTTA CREATE MORE ATP TODAY! I NEED SOME NUTRIENTS AND OXYGEN!"
OXYGEN IS ABSOLUTELY NECESSARY FOR DECOMPOSING GLUCOSE AND EXTRACTING ITS ENERGY.
HMMM...
SO TO PUT IT ANOTHER WAY, WE NEED OXYGEN TO MAKE ENERGY FROM THE FOOD WE EAT!
SO HUNGRY...
EAT DINNER
CREATE ENERGY
HEALTHY AND STRONG!
O_2
NEEDS OXYGEN

WHEN WE TAKE IN OXYGEN AND EXPEL CARBON DIOXIDE, IT'S CALLED *RESPIRATION*.
HOWEVER, THE REACTION IN WHICH CELLS ABSORB OXYGEN, BREAK DOWN GLUCOSE TO CREATE ENERGY, AND EMIT CARBON DIOXIDE IS ***ALSO*** CALLED RESPIRATION.
YOU WANT ENERGY?
GIMME SOME OXYGEN! GIMME SOME GLUCOSE!

WAIT, THEY'RE BOTH CALLED RESPIRATION? I'M GOING TO MIX THEM UP FOR SURE...

WELL, TO DISTINGUISH THEM, WE CALL THE PROCESS OF BREATHING *EXTERNAL RESPIRATION* AND THE PROCESS INSIDE OF CELLS *INTERNAL (OR CELLULAR) RESPIRATION*.
O_2
CO_2
EXTERNAL RESPIRATION
O_2
CO_2
INTERNAL RESPIRATION

THAT'S RIGHT. FROM NOW ON WHEN WE SAY "RESPIRATION," WE'LL BE REFERRING TO INTERNAL RESPIRATION.
INTERNAL! GOT THAT?

SO WE'LL BE STUDYING THE REACTION INSIDE OF CELLS, NOT THE KIND OF RESPIRATION I'M DOING RIGHT NOW.
INHALE
EXHALE

RESPIRATION IS A REACTION THAT BREAKS DOWN GLUCOSE TO CREATE ENERGY

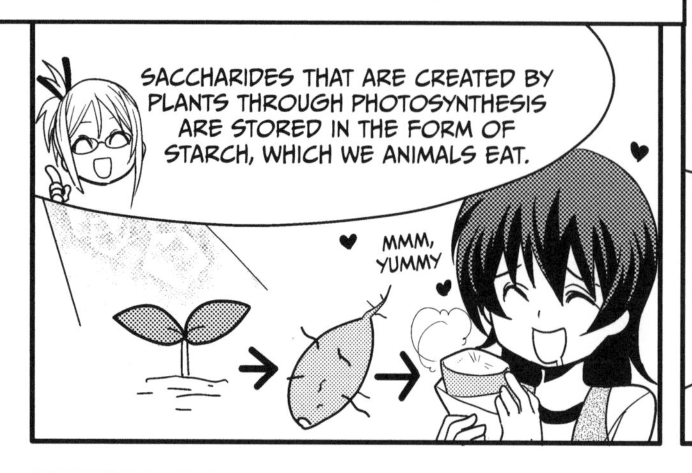

ORGANISMS LIKE US BREAK DOWN STARCH INTO GLUCOSE, WHICH IS USED TO MAKE ATP, WITH THE ASSISTANCE OF THE OXYGEN THAT WE BREATHE IN.

THAT'S INTERNAL RESPIRATION!

CO_2 ATP

$$C_6H_{12}O_6 + 6O_2 + 6H_2O \Rightarrow 6CO_2 + 12H_2O + 38ATP$$

GLUCOSE OXYGEN WATER CARBON DIOXIDE WATER ENERGY

OH, THAT MAKES IT CRYSTAL CLEAR! GLUCOSE AND OXYGEN ARE CONSUMED, AND CARBON DIOXIDE, WATER, AND ENERGY ARE PRODUCED.

NOW, LET'S TAKE A CLOSER LOOK AT THIS REACTION.

YOU CAN THINK OF RESPIRATION AS HAVING THREE STAGES...

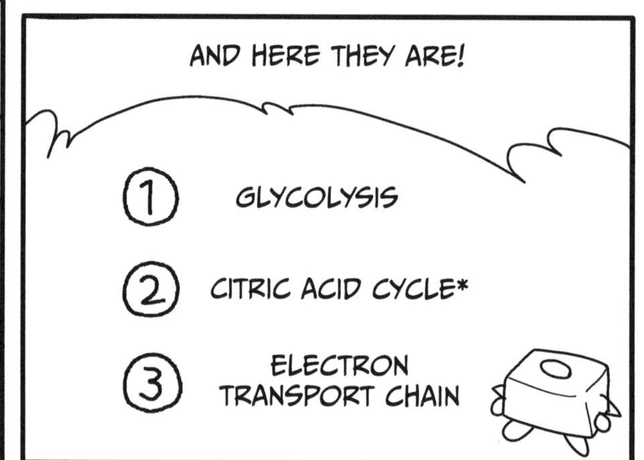

* THIS IS ALSO CALLED THE *KREBS CYCLE* OR THE *TCA CYCLE*.

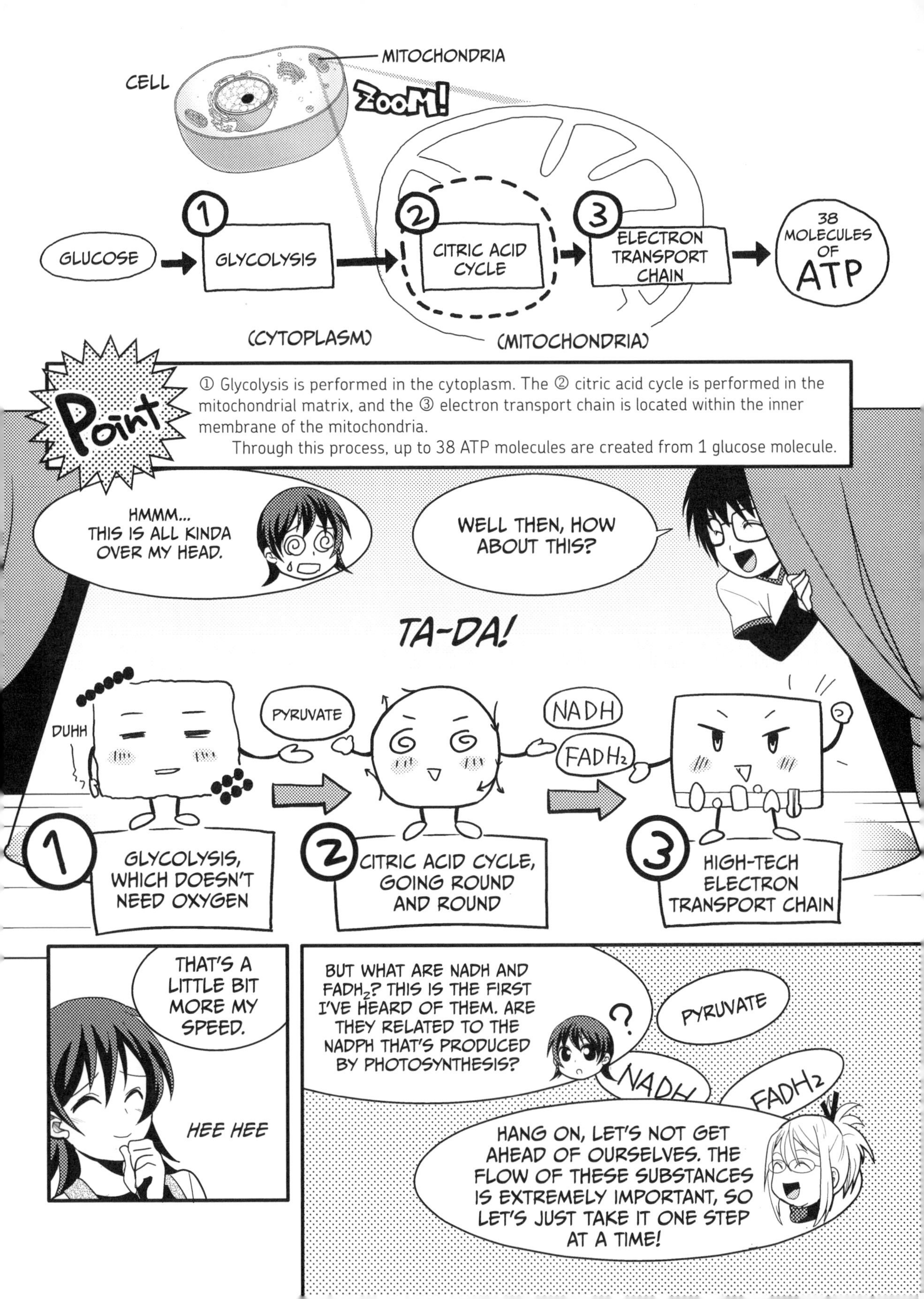
CELL
MITOCHONDRIA
ZOOM!
GLUCOSE
1 GLYCOLYSIS
2 CITRIC ACID CYCLE
3 ELECTRON TRANSPORT CHAIN
38 MOLECULES OF ATP
(CYTOPLASM)
(MITOCHONDRIA)
Point
① Glycolysis is performed in the cytoplasm. The ② citric acid cycle is performed in the mitochondrial matrix, and the ③ electron transport chain is located within the inner membrane of the mitochondria.
Through this process, up to 38 ATP molecules are created from 1 glucose molecule.
HMMM... THIS IS ALL KINDA OVER MY HEAD.
WELL THEN, HOW ABOUT THIS?
TA-DA!
DUHH
PYRUVATE
NADH
FADH2
1 GLYCOLYSIS, WHICH DOESN'T NEED OXYGEN
2 CITRIC ACID CYCLE, GOING ROUND AND ROUND
3 HIGH-TECH ELECTRON TRANSPORT CHAIN
THAT'S A LITTLE BIT MORE MY SPEED.
HEE HEE
BUT WHAT ARE NADH AND FADH2? THIS IS THE FIRST I'VE HEARD OF THEM. ARE THEY RELATED TO THE NADPH THAT'S PRODUCED BY PHOTOSYNTHESIS?
?
PYRUVATE
NADH
FADH2
HANG ON, LET'S NOT GET AHEAD OF OURSELVES. THE FLOW OF THESE SUBSTANCES IS EXTREMELY IMPORTANT, SO LET'S JUST TAKE IT ONE STEP AT A TIME!

STAGE 1: GLUCOSE DECOMPOSITION BY GLYCOLYSIS

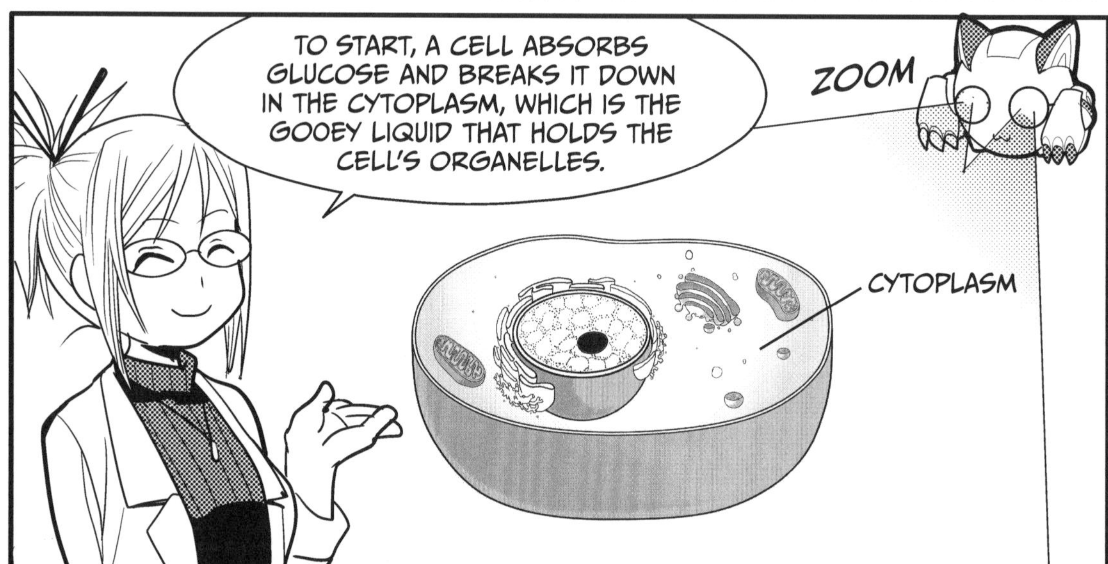

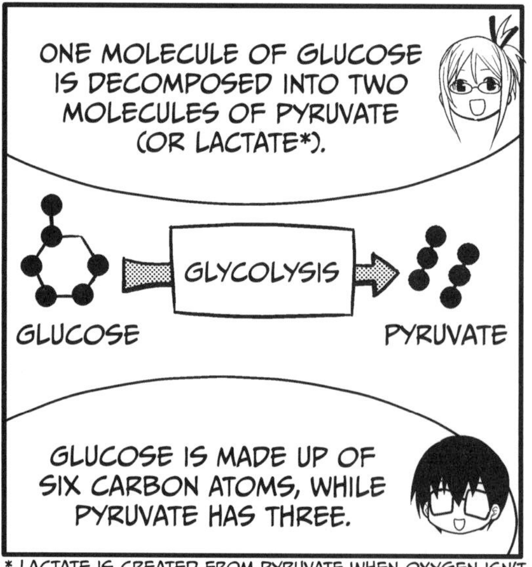

* LACTATE IS CREATED FROM PYRUVATE WHEN OXYGEN ISN'T AVAILABLE.

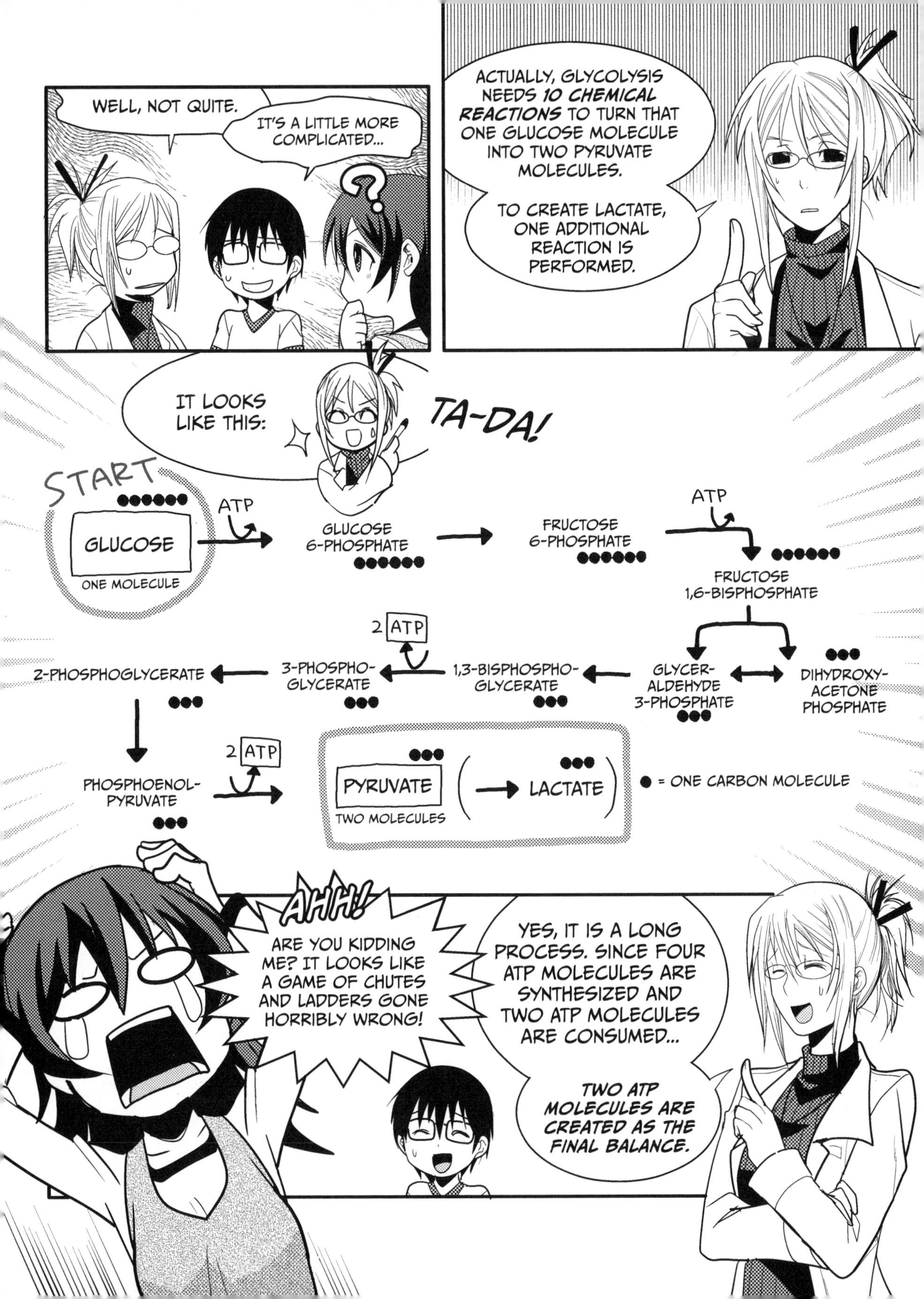
ACTUALLY, GLYCOLYSIS NEEDS 10 CHEMICAL REACTIONS TO TURN THAT ONE GLUCOSE MOLECULE INTO TWO PYRUVATE MOLECULES.
TO CREATE LACTATE, ONE ADDITIONAL REACTION IS PERFORMED.
WELL, NOT QUITE.
IT'S A LITTLE MORE COMPLICATED...
IT LOOKS LIKE THIS:
TA-DA!
START
GLUCOSE
ONE MOLECULE
ATP
GLUCOSE 6-PHOSPHATE
FRUCTOSE 6-PHOSPHATE
ATP
FRUCTOSE 1,6-BISPHOSPHATE
GLYCER-ALDEHYDE 3-PHOSPHATE
DIHYDROXY-ACETONE PHOSPHATE
1,3-BISPHOSPHO-GLYCERATE
2 ATP
3-PHOSPHO-GLYCERATE
2-PHOSPHOGLYCERATE
PHOSPHOENOL-PYRUVATE
2 ATP
PYRUVATE
TWO MOLECULES
(→ LACTATE)
● = ONE CARBON MOLECULE
YES, IT IS A LONG PROCESS. SINCE FOUR ATP MOLECULES ARE SYNTHESIZED AND TWO ATP MOLECULES ARE CONSUMED...
TWO ATP MOLECULES ARE CREATED AS THE FINAL BALANCE.
AHH!
ARE YOU KIDDING ME? IT LOOKS LIKE A GAME OF CHUTES AND LADDERS GONE HORRIBLY WRONG!

* AN ANAEROBIC ORGANISM IS ONE THAT DOESN'T REQUIRE OXYGEN FOR SURVIVAL.

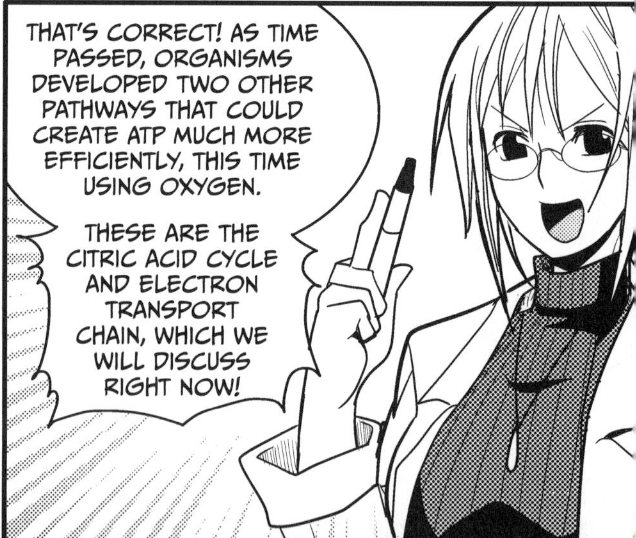

STAGE 2: CITRIC ACID CYCLE (AKA TCA CYCLE)

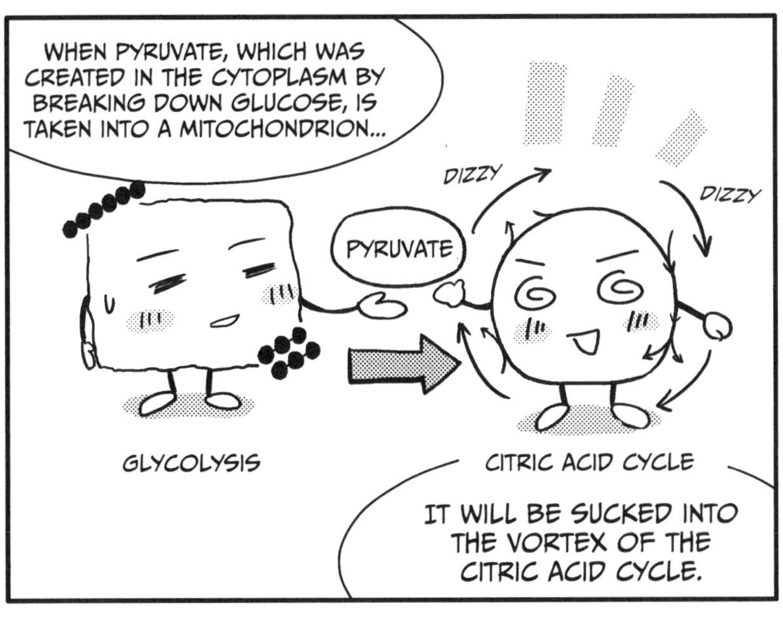

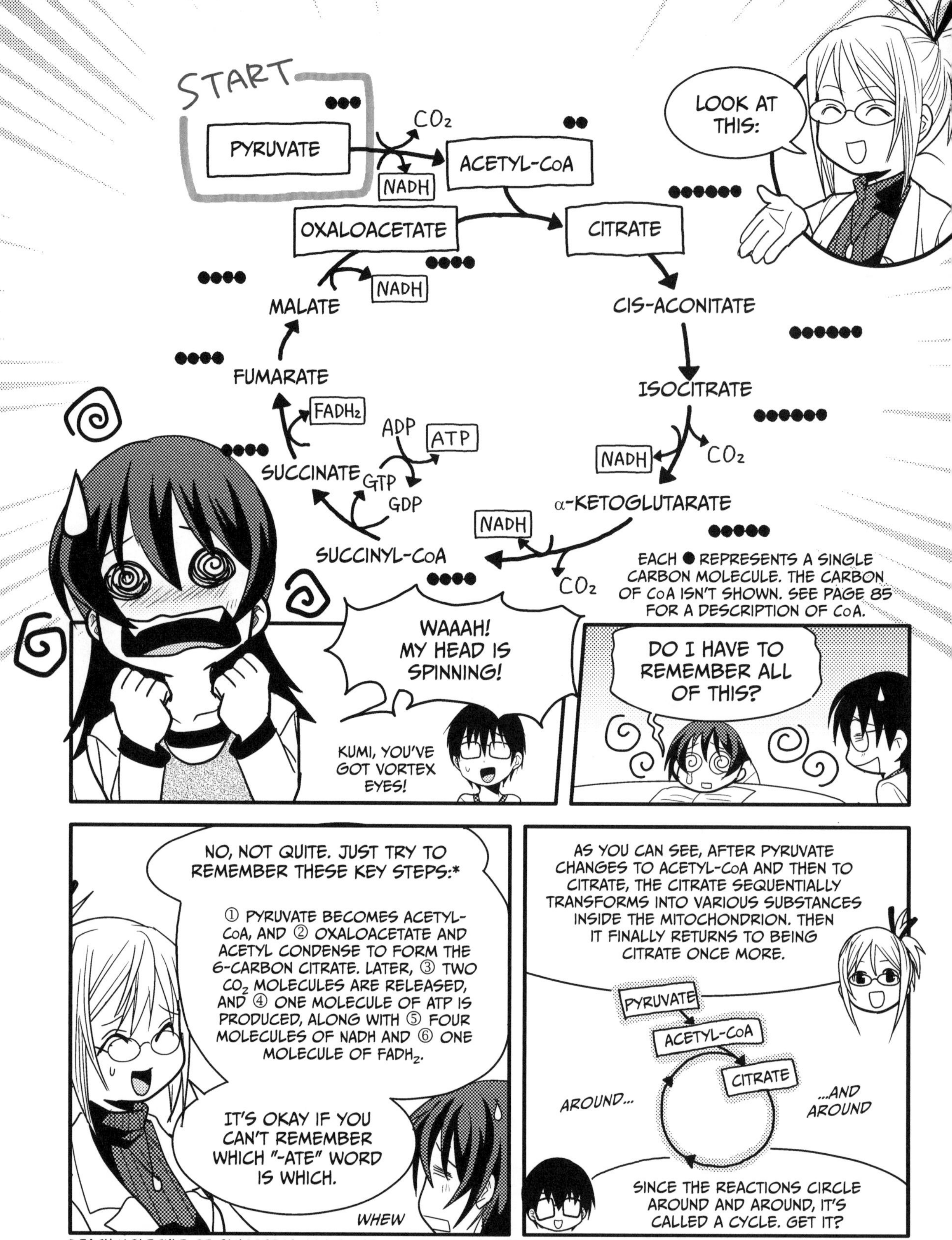

* EACH MOLECULE OF GLUCOSE IS CONVERTED INTO TWO MOLECULES OF PYRUVATE, SO TWO MOLECULES OF ATP, EIGHT MOLECULES OF NADH, AND TWO MOLECULES OF $FADH_2$ ARE CREATED IN THE CITRIC ACID CYCLE.

ALTHOUGH ONLY TWO NEW MOLECULES OF ATP ARE CREATED IN THIS CYCLE...
OTHER IMPORTANT SUBSTANCES ARE PRODUCED AS WELL.
THEY ARE THE COENZYMES NADH AND $FADH_2$!
じゃん
NADH
$FADH_2$
BY THE WAY...
WHILE NADPH IS USED IN PHOTOSYNTHESIS...
...IN RESPIRATION, IT'S JUST NADH.
I SEE...
NADPH
NADH

THESE COENZYMES CREATE A HUGE AMOUNT OF ATP IN THE ELECTRON TRANSPORT CHAIN THAT FOLLOWS.
ATP
NADH
$FADH_2$
CITRIC ACID CYCLE
ELECTRON TRANSPORT CHAIN
SO...THE COENZYMES NADH AND $FADH_2$ CONTINUE ON TO THE NEXT STAGE.
BUT WHAT DO THEY DO TO PRODUCE SO MUCH ATP?
I'LL EXPLAIN THAT NEXT!

STAGE 3: MASS PRODUCTION OF ENERGY BY THE ELECTRON TRANSPORT CHAIN

* IN OTHER WORDS, NAD^+ OR FAD IS "REDUCED," AND NADH OR $FADH_2$ IS CREATED. (SEE PAGE 37 FOR AN EXPLANATION OF REDUCTION.)

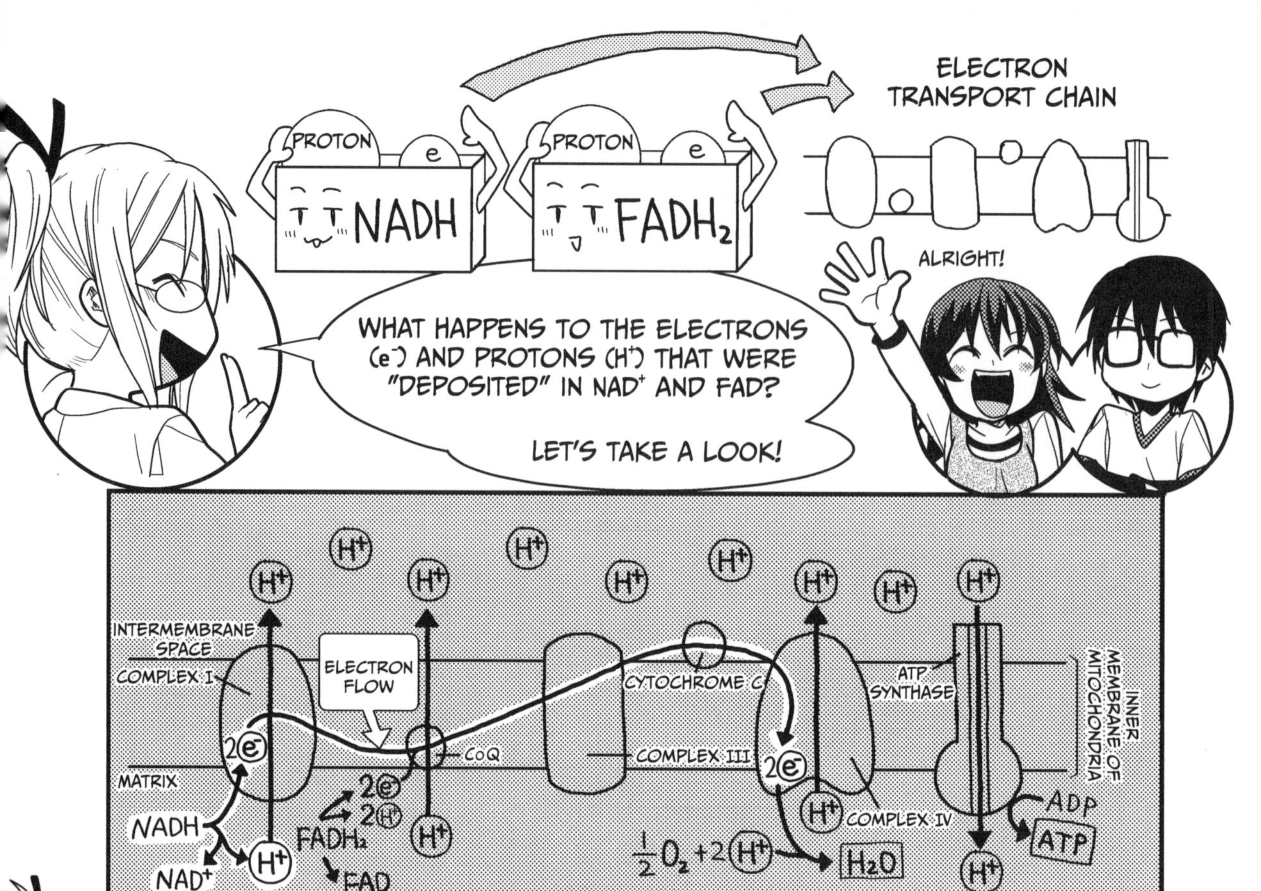

AN ELECTRON DIRECTLY ENTERS THE ELECTRON TRANSPORT CHAIN AND, AS YOU MIGHT GUESS FROM THE NAME, IS TRANSPORTED BETWEEN SEVERAL PROTEIN COMPLEXES.

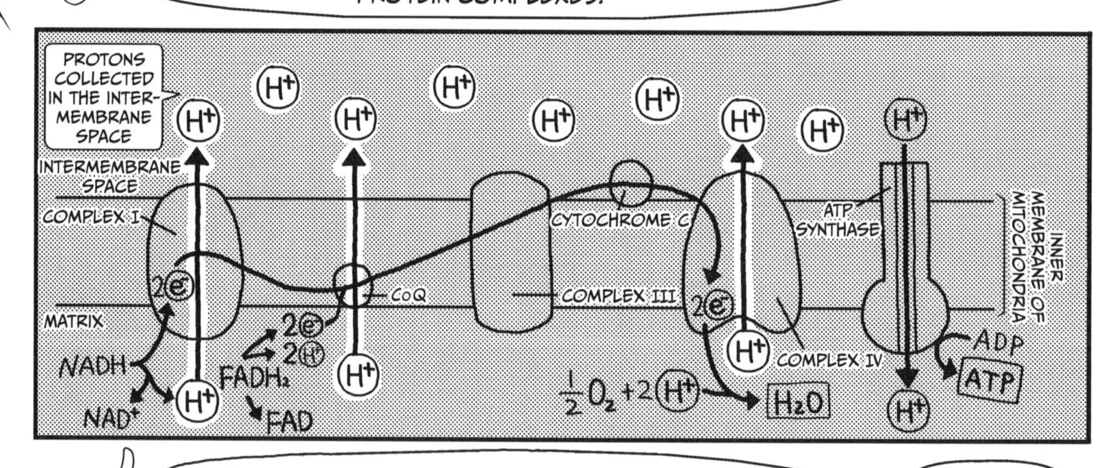

WHILE THIS ELECTRON (e^-) IS SUCCESSIVELY TRANSFERRED, THE PROTEINS UNDERGO VARIOUS CHANGES. AS A RESULT, PROTONS (H^+) MOVE FROM THE MITOCHONDRIAL MATRIX TO THE INTERMEMBRANE SPACE (BETWEEN THE INNER AND OUTER MEMBRANES). THREE "PUMPS" PUSH THESE PROTONS INTO THE INTERMEMBRANE SPACE.

WHEN THIS OCCURS, A FORCE CALLED THE *CONCENTRATION GRADIENT* IS GENERATED. THIS MEANS THAT WHEN THE CONCENTRATION OF THE PROTONS (H^+) IN THE INTERMEMBRANE SPACE BECOMES HIGHER THAN IN THE MATRIX, THE PROTONS TRY TO FLOW TOWARD THE MATRIX.

THE CONCENTRATION GRADIENT IS A FORCE WHICH CAUSES A SUBSTANCE TO NATURALLY FLOW FROM A HIGHER CONCENTRATION TO A LOWER CONCENTRATION.

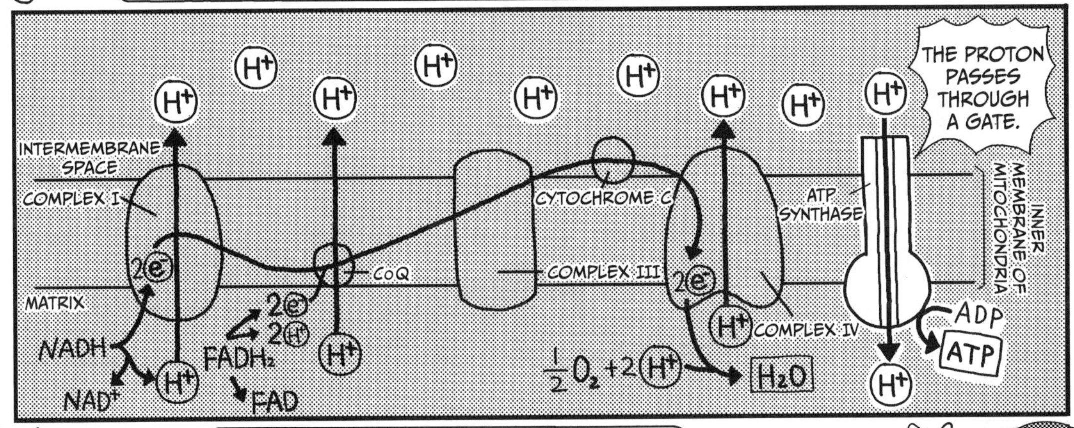

AN ELECTRON TRANSPORT CHAIN IN SATISFACTORY CONDITION HAS A "GATE" THROUGH WHICH THE PROTONS CAN MOVE INTO THE MATRIX.

OFF YOU GO, PROTONS!

THAT GATE IS THE *ATP SYNTHASE*. WHEN A PROTON PASSES THROUGH IT, ONE MOLECULE OF ATP IS PRODUCED.

THEREFORE, 30 MOLECULES OF ATP ARE CREATED FROM 10 MOLECULES OF NADH (OF WHICH TWO MOLECULES ARE PRODUCED VIA GLYCOLYSIS), AND 4 MOLECULES OF ATP ARE CREATED FROM 2 MOLECULES OF $FADH_2$.

WE GET THESE NUMBERS BECAUSE THE ELECTRONS ORIGINATING FROM NADH CAUSE THE "PROTON PUMPS" TO WORK AT ALL THREE LOCATIONS, BUT THE ELECTRONS FROM $FADH_2$ CAUSE THEM TO FUNCTION AT ONLY TWO LOCATIONS.

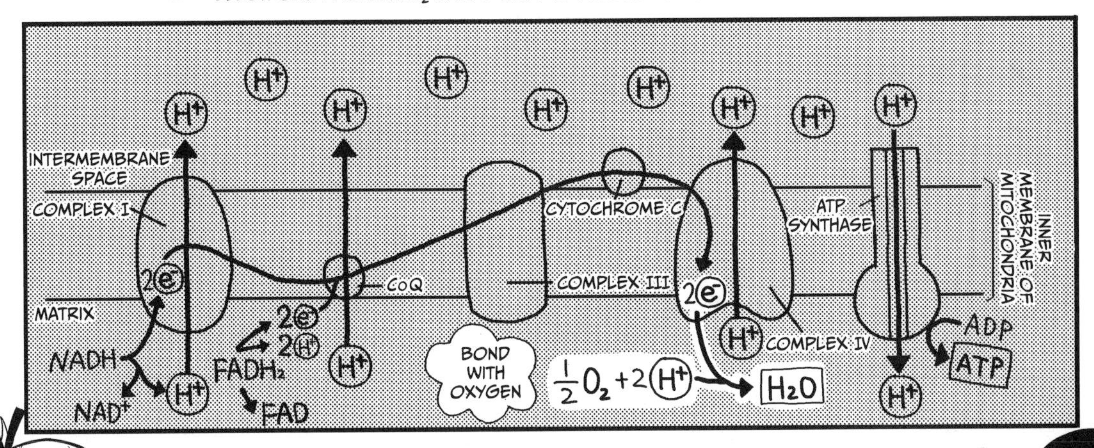

AND FINALLY THE ELECTRON AND THE PROTON THAT WERE USED BOND WITH OXYGEN (***OXYGEN IS REQUIRED HERE!***) TO PRODUCE WATER.

WE GET IT!

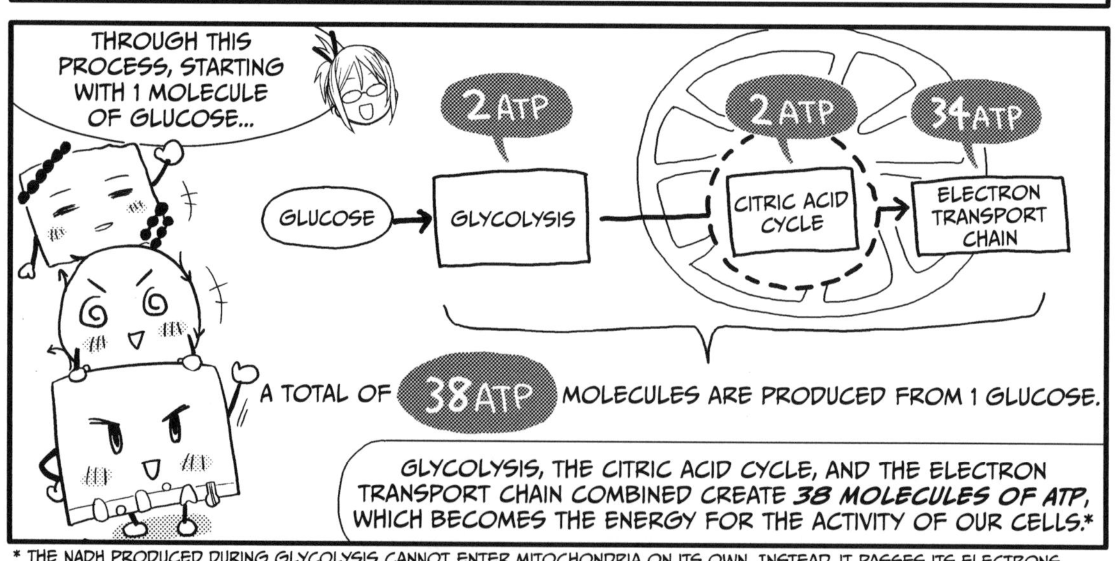

* THE NADH PRODUCED DURING GLYCOLYSIS CANNOT ENTER MITOCHONDRIA ON ITS OWN. INSTEAD, IT PASSES ITS ELECTRONS TO A "SHUTTLE" CALLED *GLYCEROL 3-PHOSPHATE*. GLYCEROL 3-PHOSPHATE PASSES ITS e^- TO FAD^+ IN THE MITOCHONDRIAL MEMBRANE, AND $FADH_2$ IS CREATED. WHEN $FADH_2$ DEPOSITS ITS e^- INTO THE ELECTRON TRANSPORT CHAIN, ONE STEP IN THE CHAIN IS SKIPPED, AND ONE LESS ATP IS FORMED (TWO LESS PER GLUCOSE), SO THE NET ATP PRODUCED IS 36.

CONCLUSION

WE CAN SUMMARIZE THE INTERRELATIONSHIPS OF PHOTOSYNTHESIS AND RESPIRATION AS FOLLOWS:
SCRIBBLE
SCRIBBLE

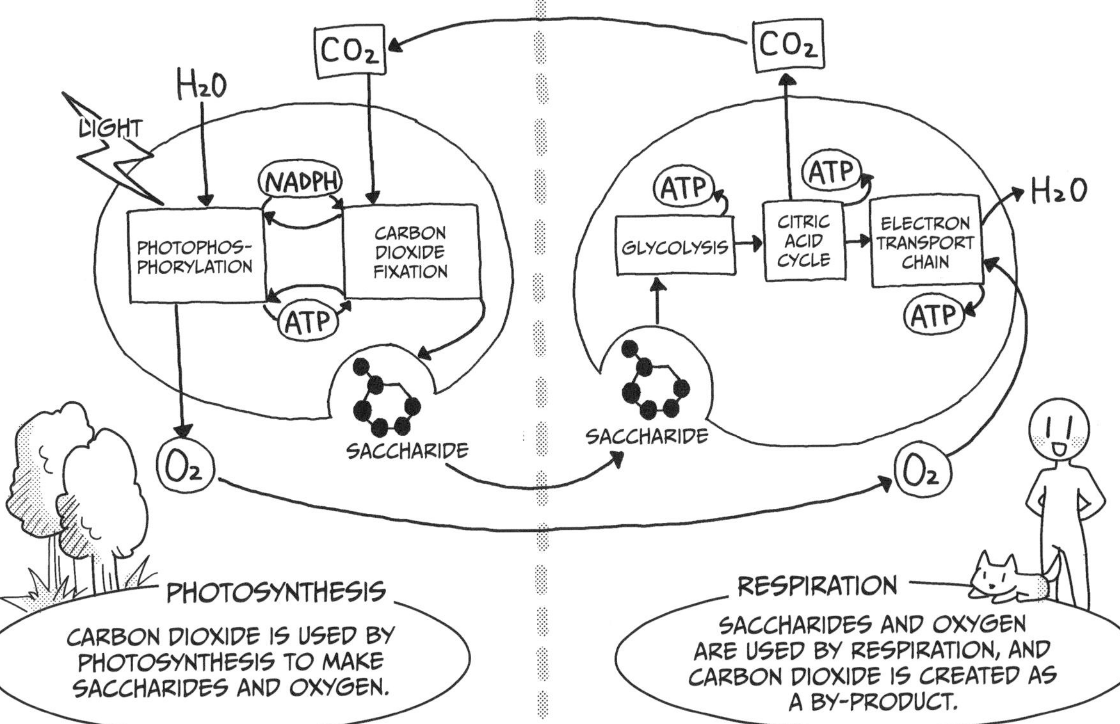
CO2
H2O
LIGHT
NADPH
PHOTOPHOS-PHORYLATION
CARBON DIOXIDE FIXATION
ATP
SACCHARIDE
O2
CO2
ATP
ATP
GLYCOLYSIS
CITRIC ACID CYCLE
ELECTRON TRANSPORT CHAIN
H2O
ATP
SACCHARIDE
O2
PHOTOSYNTHESIS
CARBON DIOXIDE IS USED BY PHOTOSYNTHESIS TO MAKE SACCHARIDES AND OXYGEN.
RESPIRATION
SACCHARIDES AND OXYGEN ARE USED BY RESPIRATION, AND CARBON DIOXIDE IS CREATED AS A BY-PRODUCT.

BY KEEPING THESE TWO PROCESSES WELL-BALANCED GLOBALLY, THE ECOSYSTEM IS MAINTAINED.
PHOTO-SYNTHESIS
RESPI-RATION

SO IF WE LOOK AT DEFORESTATION FROM A BIOCHEMICAL POINT OF VIEW, IT'S EASY TO SEE WHY IT'S SO DANGEROUS!
THE WORLDWIDE BALANCE COULD BE RUINED...

AND FARMERS WON'T BE ABLE TO GROW ANY MORE DELICIOUS FOOD.
UH, YEAH.

WOW, IT'S ALREADY DARK OUTSIDE.
YIKES!
OH MY, WE GOT SO WRAPPED UP IN STUDYING THAT WE DIDN'T EVEN NOTICE THE WHOLE DAY PASSING BY!
SCHEME!

IT'S DANGEROUS FOR A GIRL TO WALK HOME ALONE IN THE DARK...
NEMOTO, YOU SHOULD ESCORT HER HOME! RIGHT? RIGHT! THEN IT'S SETTLED!
WHAAA?!

UM...YEAH! THAT'S RIGHT! I'LL ESCORT YOU HOME! GOING OUT AT NIGHT IS SCARY AND DANGEROUS. I'LL JUST TAKE YOU BACK TO YOUR PLACE...

MY DAD IS ACTUALLY COMING TO PICK ME UP.
NO, DON'T WORRY ABOUT IT! IT'S OKAY, REALLY!
OBLIVIOUS

OH! NEMOTO, DO YOU WANT US TO GIVE YOU A RIDE?
NO THANKS... I'LL BE FINE.
SO CLOSE! WHAT A TRAGEDY...
ヨロ

I'VE GOT TO FIGURE OUT A WAY TO GET THESE TWO TOGETHER...
AHA!
ぽん

KUMI, WE SHOULD SWAP CELL PHONE NUMBERS JUST IN CASE I NEED TO CONTACT YOU FOR SOME REASON.
O-OF COURSE!

WOW, I GOT PROFESSOR KUROSAKA'S PHONE NUMBER. INCREDIBLE!
THANKS, PROFESSOR!
HE HE HE

4. ATP—The Common Currency of Energy

Plants and animals use cellular respiration to turn the sugar created by photosynthesis into potential energy, mainly in the form of adenosine triphosphate (ATP). ATP is often called the "common currency" of energy because it's used by almost every living thing: bacteria, plants—even complex organisms like Tom Cruise. However, individual molecules of ATP are *not* exchanged between organisms, so how does it operate like currency?

As you can see below, ATP has three phosphate groups attached to adenosine. When the outermost phosphate group is detached and becomes adenosine diphosphate (ADP) and inorganic phosphate (Pi), 7.3 kcal (31 kJ) per mole of energy is released. If ATP is hydrolyzed in a test tube, the surrounding water is warmed by this energy, but in an actual cell that energy is used when an enzyme catalyzes a chemical reaction, a muscle moves, or a neural signal is transmitted.

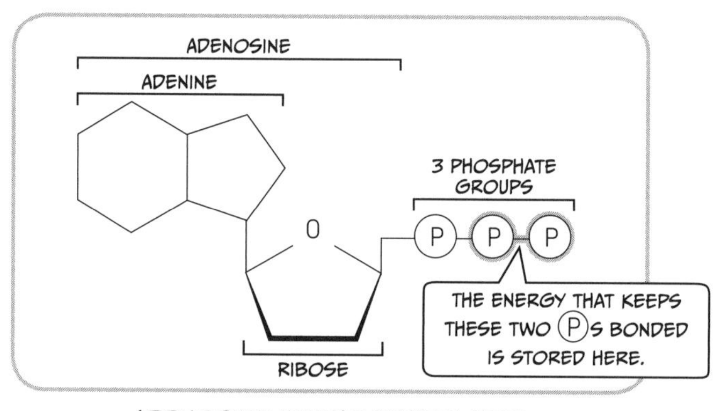

ADENOSINE TRIPHOSPHATE (ATP)

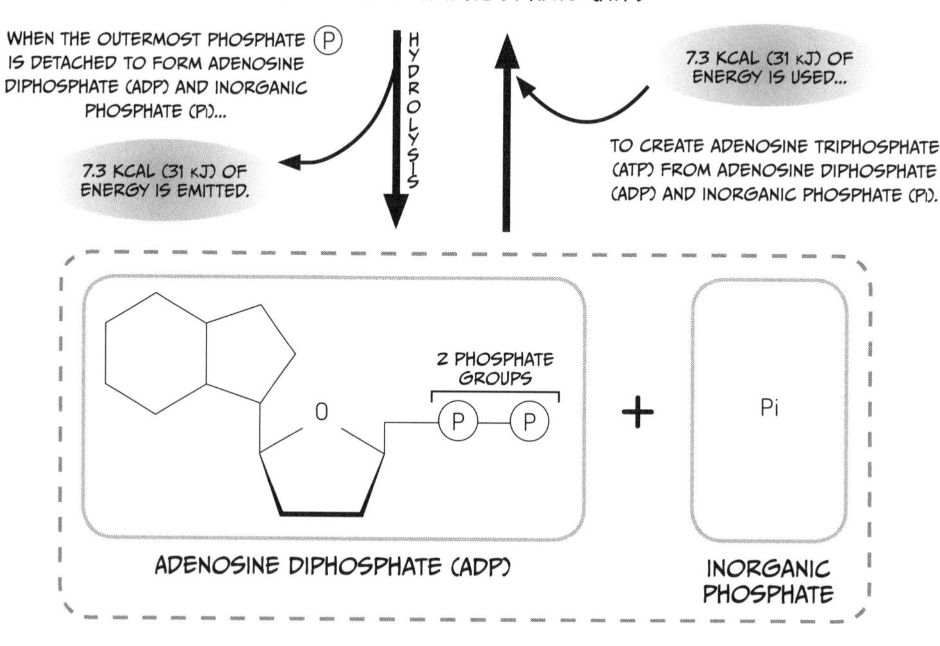

5. Types of Monosaccharides

ALDOSES AND KETOSES

We already saw that one of the basic forms of a saccharide (monosaccharide) has its first carbon forming an aldehyde group (see page 61 for more details). Another form of monosaccharide has its second carbon forming a ketone group.

Monosaccharides that have an aldehyde group are called *aldoses*, and monosaccharides with a ketone group are called *ketoses*.

Some examples of aldoses are glucose and galactose, and the most well-known ketose is fructose. (For more details about fructose, see Chapter 3.)

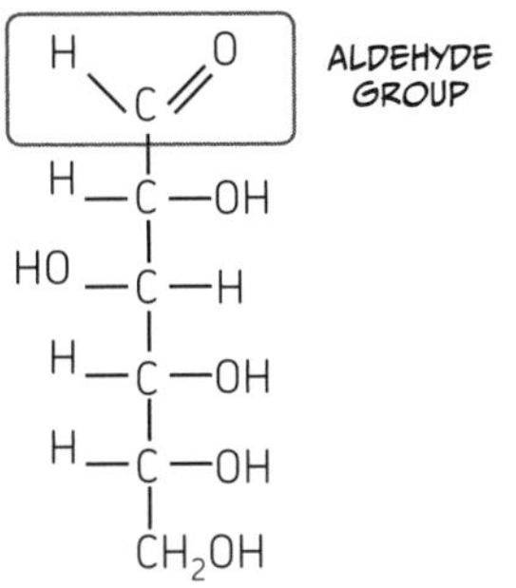

ALDEHYDE GROUP

GLUCOSE (AN ALDOSE)

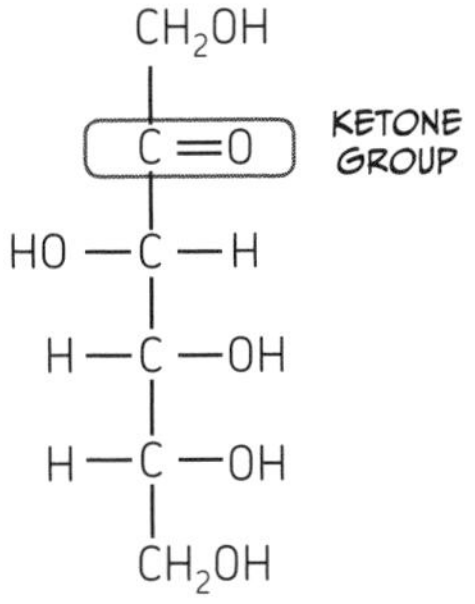

KETONE GROUP

FRUCTOSE (A KETOSE)

PYRANOSE AND FURANOSE

Earlier we learned that when certain monosaccharides, like glucose, take a cyclic structure, they resemble a hexagon. A monosaccharide that takes this form of a six-membered ring, made up of five carbons and one oxygen, is called a *pyranose*. However, there are some cases in which a monosaccharide will resemble a pentagon, made up of four carbons and one oxygen. This is called a *furanose*.

Although glucose normally takes the form of pyranose, in extremely rare cases it becomes furanose. To distinguish the two, the former is called *glucopyranose*, and the latter is called *glucofuranose*.

Fructose can also become a pyranose or furanose when it takes a cyclic structure, and these forms are called *fructopyranose* and *fructofuranose*, respectively.

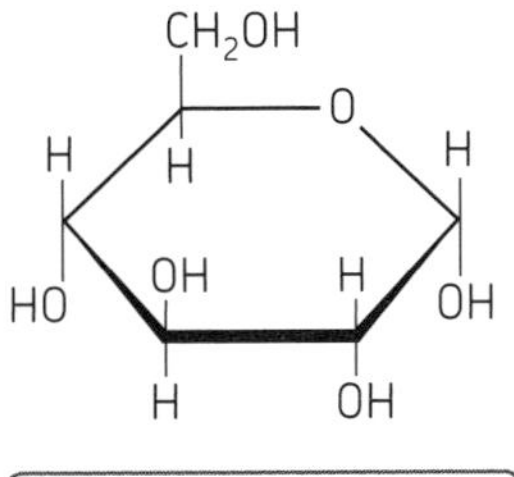

GLUCOPYRANOSE

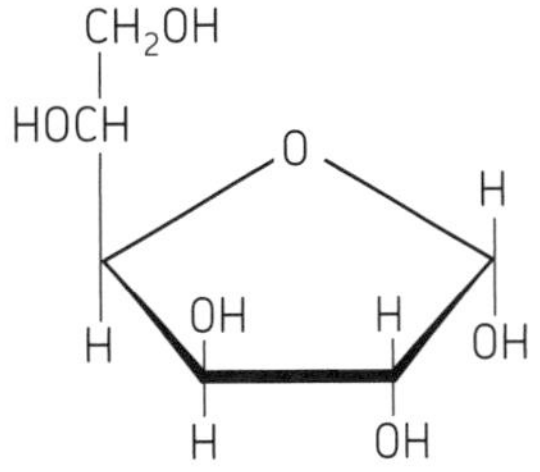

GLUCOFURANOSE

D-FORM AND L-FORM

Monosaccharides, such as glucose, can exist as *D-form* or *L-form* isomers. All monosaccharides that appear in this book are in D-form.

To tell the difference between the two forms, first find the asymmetric carbon. This is the carbon for which the four bonded substances all differ—in glucose, it's the fifth one. When the OH connected to this asymmetric carbon is on the *right* side in the structural formula of the open-chain form, this monosaccharide is the D-form. When the OH is on the *left* side, it's the L-form. Using this as a basis, the H and OH that are bonded to the carbons at the second through fifth positions of the D-form for glucose are all reversed in the L-form.

The fact that the H and OH are reversed for all carbons from the second to the fifth positions is important. If, for example, only the H and OH at the fourth position of glucose are reversed, a different monosaccharide called galactose is formed (see page 62).

Note that most monosaccharides that exist in the natural world are known to be of the D-form.

```
    H     O              H     O
     \C //                \C //
      |                    |
  H — C — OH          HO — C — H
      |                    |
 HO — C — H            H — C — OH
      |                    |
  H — C — OH          HO — C — H
      |                    |
 [H — C — OH] D-FORM  [HO — C — H] L-FORM
      |                    |
     CH2OH                CH2OH

   D-GLUCOSE            L-GLUCOSE
```

6. What Is CoA?

When pyruvate, which is created by glycolysis, enters into the citric acid cycle, it becomes a substance called acetyl-CoA. But what does CoA mean?

CoA stands for *coenzyme A*. Its structural formula is shown below. CoA is a substance in which two phosphates in a row are bonded to the fifth carbon of adenosine triphosphate, and a vitamin called pantothenic acid, as well as 2-mercaptoethylamine, are also bonded. Within the CoA, shown in the shaded part of the figure below, is the *phosphopantetheine group*, which works as a "carrier" because it transports the acetyl group (aka the carbohydrate chain of fatty acids). Acetyl-CoA has an acetyl group bonded at the front of this intimidating-looking molecule.

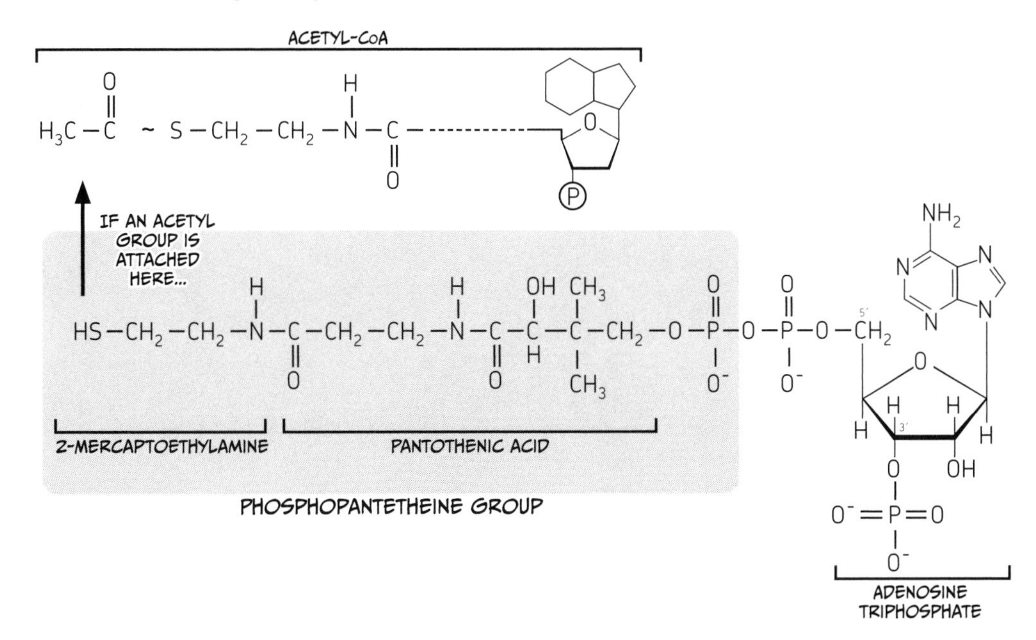

STRUCTURE OF CoA

Another protein that works similarly to CoA is ACP (acylcarrier protein). ACP, which we'll examine more thoroughly in Chapter 3, is also a "carrier." And like CoA, it has a phosphopantetheine group but at a different location. The phosphopantetheine group is bonded to the serine (a type of amino acid) of ACP rather than to adenosine triphosphate.

Since it's called a coenzyme, CoA plays the role of assisting in chemical reactions necessary in the procession of a metabolic pathway.

BIOCHEMISTRY IN OUR EVERYDAY LIVES

1. Lipids and Cholesterol

TWO HOURS LATER—

HELLO

I WONDER WHAT HAPPENED TO THE PROFESSOR SO SUDDENLY. I HOPE I CAN TEACH THIS STUFF PROPERLY...

FIRST UP:
1. IS CHOLESTEROL REALLY BAD?
LET'S FIND OUT.
YEAH!
CHOLESTEROL IS OIL OR FAT, RIGHT? I THINK IT'S DEFINITELY A BAD THING.
DAD
CHOLESTEROL = BAD!
I KNOW MY DAD IS ALWAYS GOING ON AND ON ABOUT HIS CHOLESTEROL LEVELS...
FAT IS UNHEALTHY, AND WE DON'T NEED IT. MY GOAL IS ZERO PERCENT BODY FAT!
GOAL: LOSE 5 LBS!
DOWN WITH THE POUNDS!
THAT'S NOT REALLY TRUE, KUMI.
FOR YOUR SAKE, WE'D BETTER STUDY LIPIDS NEXT. ALTHOUGH WE LEARNED ABOUT SACCHARIDES IN PROFESSOR KUROSAKA'S LESSONS, WE ALSO NEED TO LEARN ABOUT TWO MORE THINGS: LIPIDS AND PROTEINS. THESE ARE THE THREE MAJOR NUTRIENTS. WELL THEN, LET'S TALK ABOUT LIPIDS!
AHEM
WHAT? I'M JUST WORRIED ABOUT FAT. WHAT ARE LIPIDS? ARE THEY DIFFERENT FROM FAT?

* THERE ARE EXCEPTIONS, THOUGH: SOME GLYCOLIPIDS DO DISSOLVE IN WATER.

NOW LET'S DISCUSS THESE DIFFERENT LIPID TYPES, ONE BY ONE.
NEUTRAL FAT
• NEUTRAL LIPID
FIRST, WE HAVE *NEUTRAL LIPIDS!*
GRRRRRRRRRR
AS I MENTIONED EARLIER, A NEUTRAL LIPID IS THE SUBSTANCE THAT WE NORMALLY CALL "FAT."
FAT
NEUTRAL LIPIDS ARE FORMED FROM TWO SUBSTANCES: GLYCEROL AND FATTY ACIDS.
AMONG THE NEUTRAL LIPIDS INSIDE OUR BODIES, THE MOST COMMON IS TRIACYLGLYCEROL, WHICH CONSISTS OF ONE GLYCEROL MOLECULE AND THREE FATTY ACIDS, COMBINED LIKE THIS:

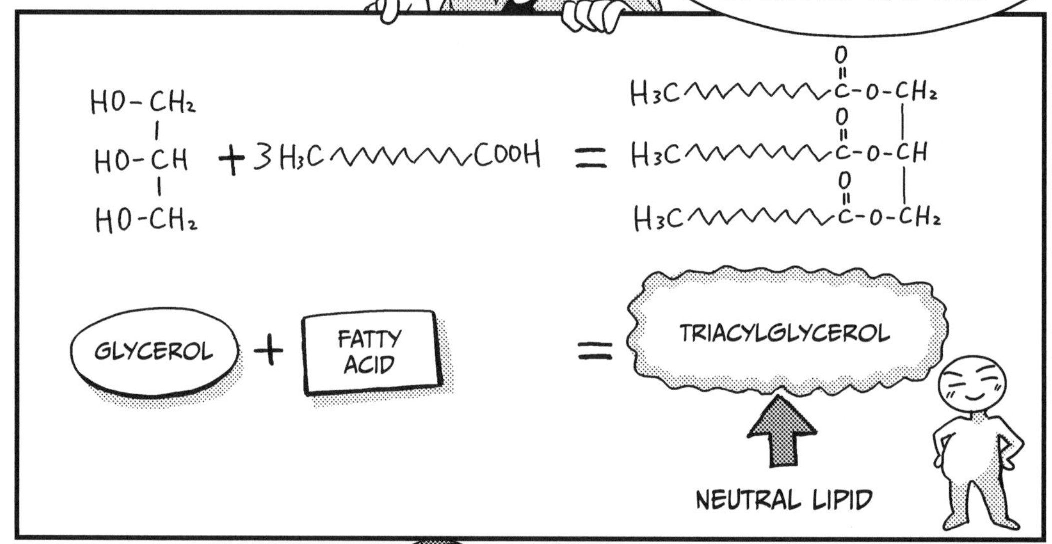
HO−CH₂
HO−CH
HO−CH₂
+ 3H₃C∿∿∿COOH =
H₃C∿∿∿C(=O)−O−CH₂
H₃C∿∿∿C(=O)−O−CH
H₃C∿∿∿C(=O)−O−CH₂
GLYCEROL + FATTY ACID = TRIACYLGLYCEROL
NEUTRAL LIPID

TRIACYLGLYCEROL WILL OFTEN POP UP IN DISCUSSIONS ABOUT LIPIDS.
SHAKE
AHA! MY ENEMY IS TRIACYLGLYCEROL! I'LL NEVER FORGET THE NAME OF THIS ABOMINABLE FOE!
THERE'S ALSO MONOACYLGLYCEROL, WHICH HAS JUST ONE FATTY ACID COMBINED WITH ONE GLYCEROL...
AND DIACYLGLYCEROL, IN WHICH TWO FATTY ACIDS ARE COMBINED WITH ONE GLYCEROL.
MOVING RIGHT ALONG...

NEXT UP ARE PHOSPHOLIPIDS!
• PHOSPHOLIPID

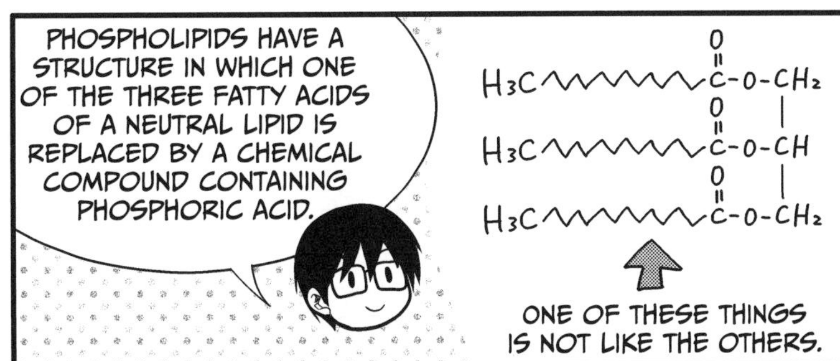
PHOSPHOLIPIDS HAVE A STRUCTURE IN WHICH ONE OF THE THREE FATTY ACIDS OF A NEUTRAL LIPID IS REPLACED BY A CHEMICAL COMPOUND CONTAINING PHOSPHORIC ACID.
H_3C — $C(=O)$ — O — CH_2
H_3C — $C(=O)$ — O — CH
H_3C — $C(=O)$ — O — CH_2
ONE OF THESE THINGS IS NOT LIKE THE OTHERS.

BY THE WAY, KUMI, WE ALREADY TALKED ABOUT PHOSPHOLIPIDS ONCE, REMEMBER?
UMM...NOT REALLY...
GUUUH
REMEMBER WHEN WE TALKED ABOUT THE *CELL MEMBRANE*?
IT CONSISTS MAINLY OF PHOSPHOLIPIDS!
ZOOM!
PHOSPHOLIPIDS HAVE A PROPERTY CALLED *AMPHIPATHICITY*.
THE TWO FATTY ACID PARTS ARE HYDROPHOBIC, WHILE THE CHEMICAL COMPOUND PART CONTAINING PHOSPHORIC ACID IS HYDROPHILIC.
BECAUSE OF THIS, A TWO-LAYER MEMBRANE CAN BE CREATED WITH THE HYDROPHOBIC PART FACING INWARD AND THE HYDROPHILIC PART FACING OUTWARD.
PHOSPHOLIPID
PHOSPHORIC ACID → HYDROPHILIC
FATTY ACID → HYDROPHOBIC
PHOSPHO-LIPID
HYDROPHILIC
H_3C — $C(=O)$ — O — CH_2
H_3C — $C(=O)$ — O — CH
H_2C — O — $P(=O)(O^-)$ — □
HYDROPHOBIC
VARIABLE POLAR HEAD*
* A PHOSPHOLIPID CAN HAVE A VARIABLE POLAR GROUP AS ITS HEAD. A PHOSPHOLIPID THAT IS BASED ON GLYCEROL IS CALLED A *GLYCEROPHOSPHOLIPID*. *SPHINGOPHOSPHOLIPIDS* HAVE SPHINGOSINE HEADS.
HYDROPHILIC MEANS IT READILY MIXES WITH WATER. HYDROPHOBIC MEANS IT DOES NOT READILY MIX WITH WATER. AMPHIPATHIC MEANS THAT IT'S MADE FROM BOTH OF THESE KINDS OF SUBSTANCES.

AND NOW, GLYCOLIPIDS!
• GLYCOLIPID
GLYCOLIPIDS ARE LIPIDS THAT CONTAIN A SACCHARIDE AS A COMPONENT.
THERE ARE VARIOUS TYPES OF GLYCOLIPIDS, SUCH AS SPHINGOGLYCOLIPIDS AND GLYCEROGLYCOLIPIDS.
SACCHA-RIDE
GALACTOCEREBROSIDE (A TYPE OF SPHINGOGLYCOLIPID)
CH_2OH
GALACTOSE
SPHINGOSINE
FATTY ACID
$(CH_2)_{12}$
CH_3
PHOSPHOLIPIDS AND GLYCOLIPIDS ALSO INCLUDE FATTY ACIDS, JUST LIKE NEUTRAL LIPIDS DO.
LIPIDS
NEUTRAL FAT
• NEUTRAL LIPID
• PHOSPHOLIPID
• GLYCOLIPID
• STEROID
THESE CONTAIN FATTY ACIDS.
HMM...FATTY ACIDS ARE CONTAINED IN A BUNCH OF DIFFERENT LIPIDS, AREN'T THEY?
THAT'S RIGHT! MOST LIPIDS CONTAIN FATTY ACIDS.
THEY'RE THE STARS OF THE SHOW HERE, ACTUALLY.
STARS?! YEAH RIGHT! I'D LIKE TO PUT THEIR LIGHTS OUT...
KUMI, THE FATTY ACIDS YOU HATE SO MUCH ARE ACTUALLY VERY IMPORTANT.
LET'S TAKE A CLOSER LOOK.

FATTY ACIDS

Fatty acids are a source of energy, and they can become phospholipids, which are a raw material for creating cell membranes. If there were no fatty acids, humans could not live.

Wow, really? That's surprising! And I thought they were the enemy.

First, let's look at the structure of fatty acids. Although they can be constructed by connecting a few to several dozen carbon (C) atoms together, the fatty acids that are in our bodies always contain 12 to 20 carbon atoms.

$(CH_3(CH_2)_{12}COOH)$

CARBOXYL GROUP

At the farthest end of that long chain (which is called a *hydrocarbon chain*) is a structure called a *carboxyl group* (-COOH).

Since only hydrogen (H) atoms are attached to each of the carbon atoms, the fatty acid does **not** mix easily with water. It lacks hydroxyl groups (OH), which saccharides have (see page 61 for information about saccharides).

Oh, I get it, like oil and water. So tough to mix!

Some fatty acids are made in our bodies. For instance, excess carbohydrates are converted into palmitic acid. Two fatty acids, linoleic and α-linolenic acid, are *essential*, which means that they are necessary for good health and cannot be synthesized by humans. On the other hand, stearic acid and arachidonic acid cannot be synthesized but are not essential to good health either. All these fatty acids contain over 16 Cs!

Yikes! So many Cs...it's like a nightmare report card!

Palmitic acid	$CH_3(CH_2)_{14}COOH$
Stearic acid	$CH_3(CH_2)_{16}COOH$
Linoleic acid	$CH_3(CH_2)_4(CH{=}CHCH_2)_2(CH_2)_6COOH$
α-linolenic acid	$CH_3CH_2(CH{=}CHCH_2)_3(CH_2)_6COOH$
Arachidonic acid	$CH_3(CH_2)_4(CH{=}CHCH_2)_4(CH_2)_2COOH$

When there is a double bond inside the molecule, we write it like this.

Certain fatty acids have double bonds between the carbon atoms in the middle of the molecule, as you can see in this figure.

A CARBON ATOM HAS FOUR "ARMS," AND USUALLY A SEPARATE ATOM BONDS WITH EACH ARM.

HOWEVER, IN SOME CASES, TWO ARMS ARE USED TO BOND A CARBON WITH ANOTHER ATOM. THIS IS CALLED A *DOUBLE BOND*.

Carbons that are double bonded are called *unsaturated carbons*, and fatty acids that have unsaturated carbons are called *unsaturated fatty acids*.

Unsaturated fatty acids don't solidify as easily, remaining as liquid at lower temperatures than saturated fatty acids, so they are often included as a component of cell membranes (that is, phospholipids) for which flexibility is important.

So...if there are a lot of double bonds, fatty acids are harder to solidify?

That's right. Double bonds can create kinks, which prevent the unsaturated fatty acids from forming a stable solid. This means that **the melting point of fatty acids differs significantly depending on the number of carbon atoms and the number of double bonds between those carbon atoms**.

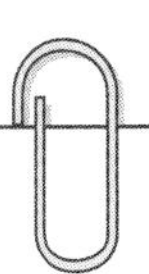

 CHOLESTEROL IS A TYPE OF STEROID

Okay, before we get too off track, let's get back to cholesterol.

Yeah, that's right! I'm pretty sure I understand lipids and fatty acids, but what exactly is cholesterol?

Cholesterol is also a type of lipid, but it has the following form:

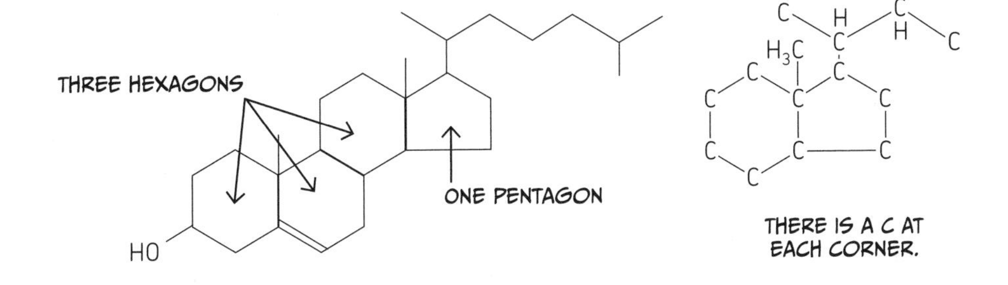

The three hexagons and one pentagon combined, as shown in the above figure, are called a *steroidal skeleton*, and a lipid that has this basic form is called a *steroid*.

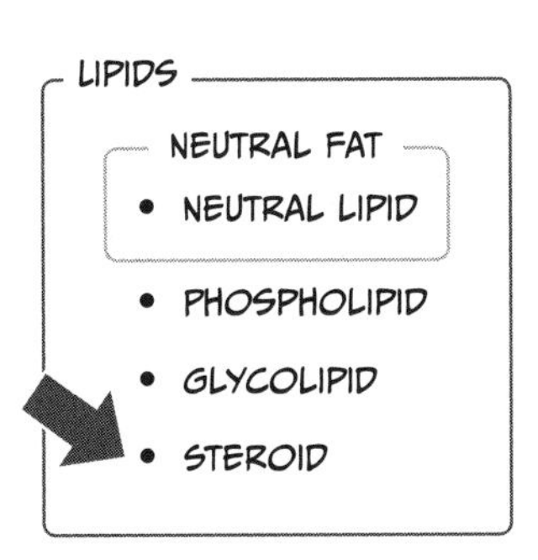

Well...I guess that means that steroids are a type of fat! Weird!

CHOLESTEROL'S JOB

I understand that cholesterol is a type of steroid, but what exactly **is** a steroid? All I know is that bodybuilders use them to get super buff.

Well, when we talk about "steroids," it often brings to mind images of pharmaceuticals and intimidating pectoral muscles, but there are actually a number of steroids that exist in our bodies normally, like cholesterol.

For example, there is a type of hormone called a *steroid hormone*. The most famous of these are the sex hormones—the indispensible hormones that make males males and females females.

Sex hormones, eh? Whatever they are, I'm just glad they made me a girl. Boys are totally gross!

Um...whatever you say, Kumi.

Anyway, *testosterone*, which is a sex hormone produced mainly in a male's testicles, is actually created with cholesterol as a raw material.

Progesterone, another sex hormone, is produced in a female's ovaries or placenta and is also created from cholesterol.

OH

O

TESTOSTERONE

CH_3

C=O

O

PROGESTERONE

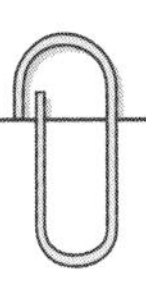

Vitamin D is also a type of steroid created from cholesterol. Since Vitamin D is produced when ultraviolet rays strike our skin, it's very important for human beings to be exposed to the sun.

ULTRAVIOLET (UV) RAY

CHOLESTEROL ⟶ 7-DEHYDROCHOLESTEROL ⟶

CH_2

HO OH

VITAMIN D_3

In addition, cholesterol plays other very important roles, such as being a raw material of bile acid,* which is required for digestion and the absorption of fat in the small intestine.

Wow! Cholesterol has a ton of important jobs! I had no idea...

That's right. When people who pay attention to their health (like your father) talk about cholesterol, usually a negative image, like arteriosclerosis or obesity, comes to mind, but cholesterol is actually an extremely important substance for our bodies.

Jeez, now I'm really confused. Why does such a vital substance make us think of illness, unhealthiness, or being overweight?

Don't worry, we'll unravel that mystery soon enough...

* Bile acid is created in the liver, stored in the gallbladder, and secreted to the duodenum.

LIPOPROTEINS: BEYOND GOOD AND EVIL

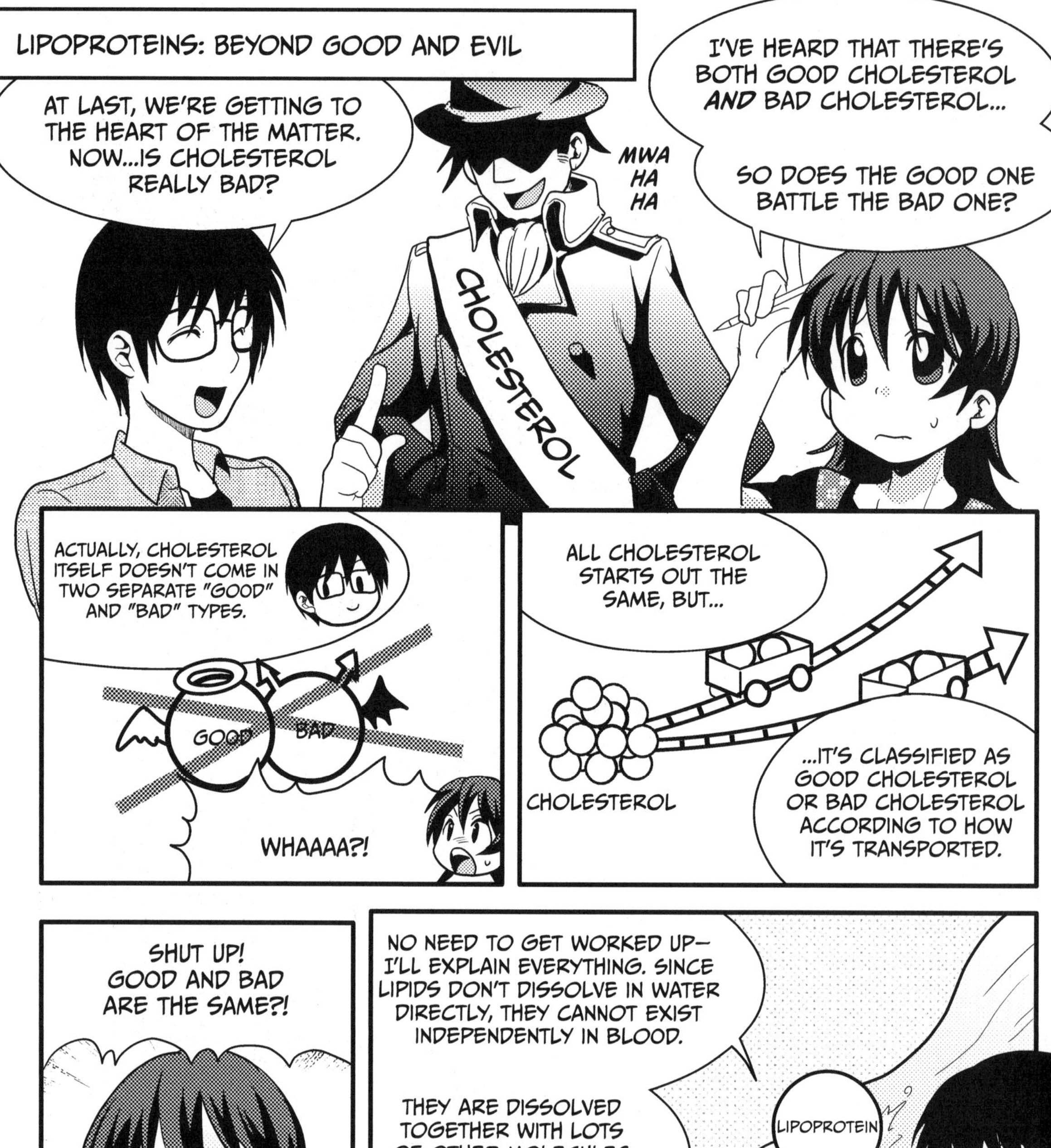

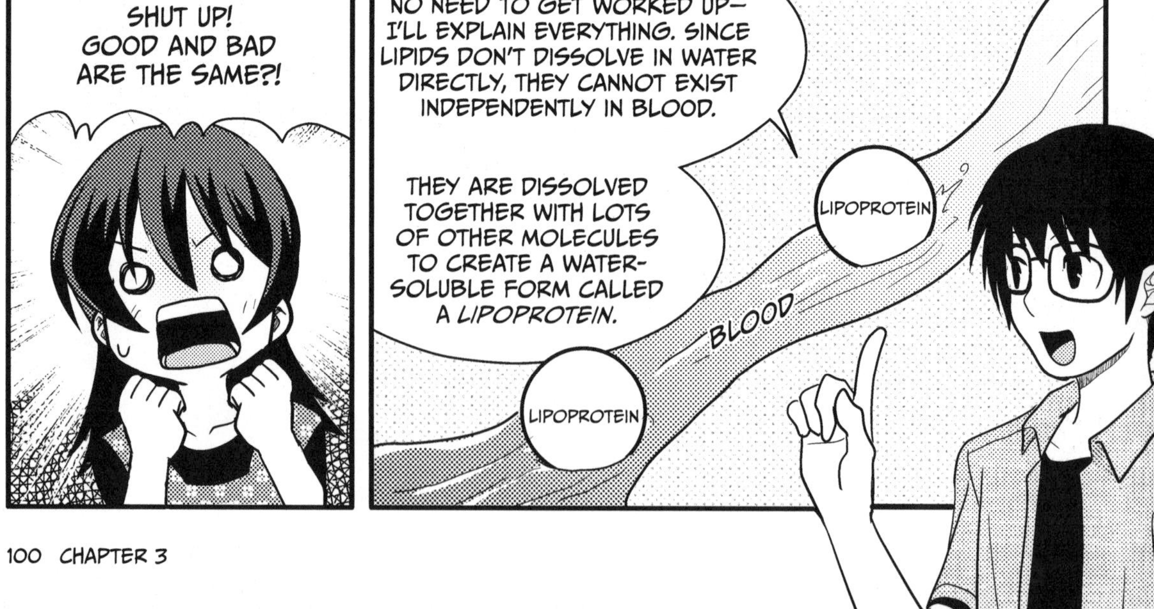

* CHOLESTEROL ESTER IS CHOLESTEROL WITH A FATTY ACID ATTACHED TO IT.

** A FREE CHOLESTEROL IS A CHOLESTEROL WITHOUT A FATTY ACID ATTACHED TO IT.

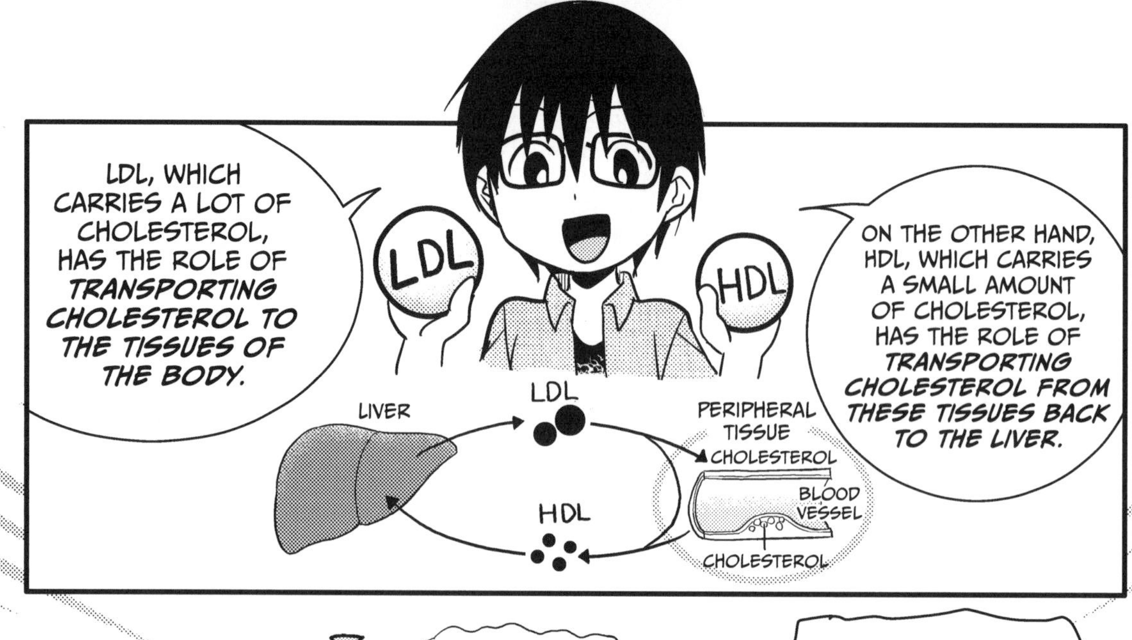
LDL, WHICH CARRIES A LOT OF CHOLESTEROL, HAS THE ROLE OF *TRANSPORTING CHOLESTEROL TO THE TISSUES OF THE BODY.*
LDL
HDL
ON THE OTHER HAND, HDL, WHICH CARRIES A SMALL AMOUNT OF CHOLESTEROL, HAS THE ROLE OF *TRANSPORTING CHOLESTEROL FROM THESE TISSUES BACK TO THE LIVER.*
LIVER
LDL
PERIPHERAL TISSUE
CHOLESTEROL
BLOOD VESSEL
HDL
CHOLESTEROL

HEHE, YOU CAN NEVER HAVE ENOUGH CHOLESTEROL!
BAD
LDL
OH NO...A BATTLE OF GOOD VERSUS EVIL!
LIVER
BODY TISSUES
ACK!
I KEEP THINGS TIDY BY BRINGING THE EXCESS BACK TO THE LIVER.
GOOD
HDL
SO LDL IS CALLED "BAD" BECAUSE IT DISTRIBUTES CHOLESTEROL THROUGHOUT THE BODY, AND HDL IS CALLED "GOOD" BECAUSE IT REMOVES CHOLESTEROL FROM PERIPHERAL TISSUES AND BRINGS IT BACK TO THE LIVER.
GOOD
BAD
I FINALLY GET IT! THE CHOLESTEROL ITSELF IS ALL THE SAME.
SMACK
BUT IT'S REGARDED AS "GOOD" OR "BAD" BASED ON THE TYPE OF LIPOPROTEIN THAT'S TRANSPORTING IT!
GOOD
BAD

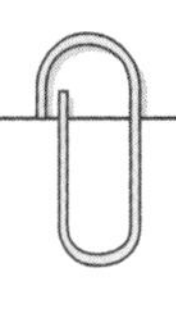

WHAT IS ARTERIOSCLEROSIS?

Now that you have a better understanding of cholesterol, imagine what would happen if the density of LDL (bad) in the blood increases while the density of HDL (good) decreases.

Hmmm, I guess cholesterol would be transported into the body's tissues faster than it could be removed, and it would pile up in the blood vessels.

Right. Cholesterol will accumulate on the walls of blood vessels, the lumens of the blood vessels will get narrower, and the flow of blood will be obstructed. This is called *arteriosclerosis*.

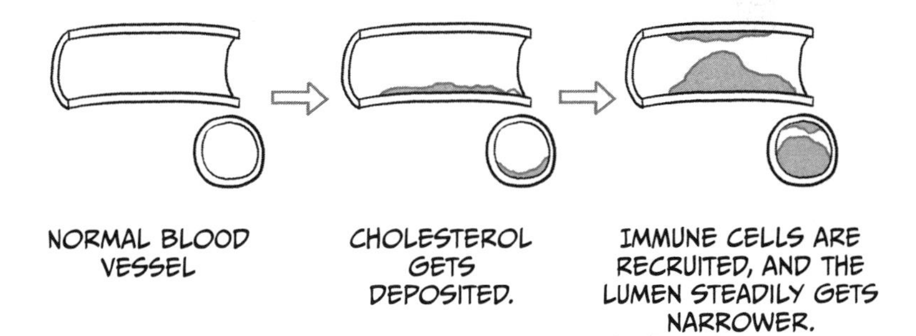

The blood vessels get thicker and harder, and as the symptoms advance, it can lead to sickness or even death.

Cholesterol can lead to **death**?! Not to be taken lightly, I guess...

Let's look at a typical type of arteriosclerosis called *atherosclerosis*. Scientists believe the process goes something like this: First, LDL cholesterol is deposited on the inner wall of a damaged blood vessel. This cholesterol is then eaten by phagocytes,* such as *macrophages*, and bulging cells filled with fat, called *foam cells*, accumulate.

When this happens, the smooth muscle cells that create the walls of a blood vessel also end up changing significantly, getting harder and thicker.

* A *phagocyte* is a gluttonous kind of white blood cell that eats just about anything. They're an essential part of the immune system.

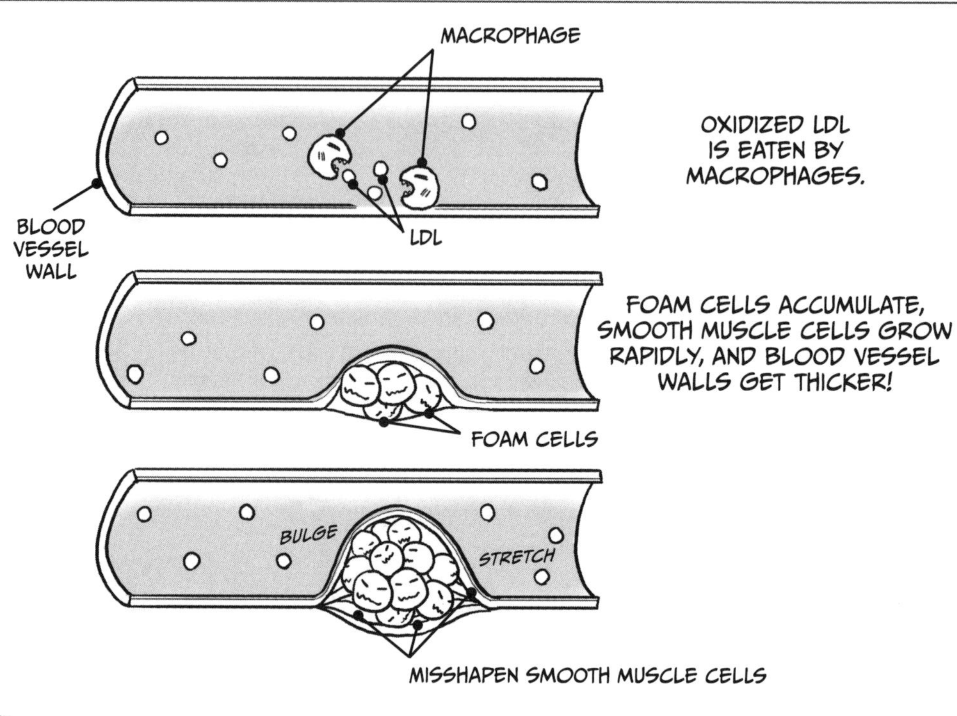

Wow, that artery is even more clogged than the pipes in my bathtub. But at least some hair in the drain won't kill me!

Don't be so sure. The clog in my tub practically has a mind of its own. It's a real jungle down there!

Whoa, gross! TMI, Nemoto!

Er...sorry. Anyway, you can see some of the illnesses that arteriosclerosis can cause below.

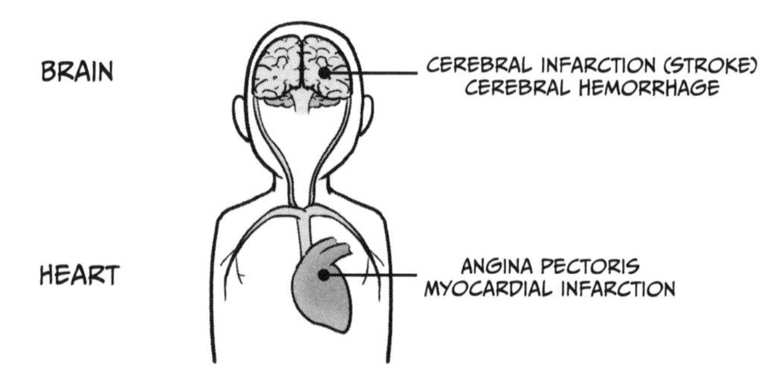

IS CHOLESTEROL REALLY BAD?

- Cholesterol, which is transported through the blood via lipoproteins, can be both good (HDL) and bad (LDL).
- The cholesterol in LDL is transported to the body's peripheral tissues, and the cholesterol in HDL is transported from peripheral tissues to the liver. The balance of these is very important.
- Cholesterol is an important substance that makes hormones, but if too much is absorbed, it can result in arteriosclerosis, which can lead to serious illnesses and even death.

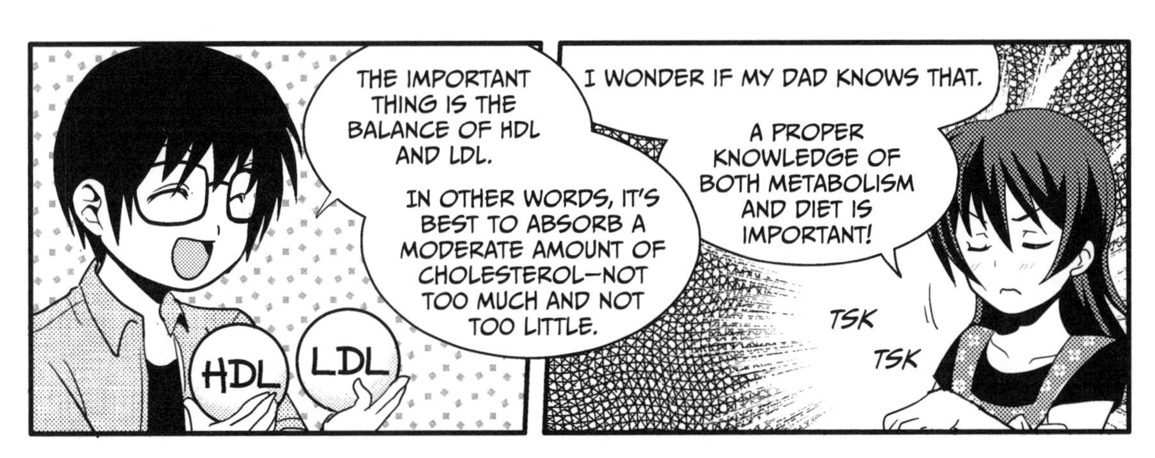

2. Biochemistry of Obesity—Why Is Fat Stored?

INGESTED AND EXPENDED ENERGY

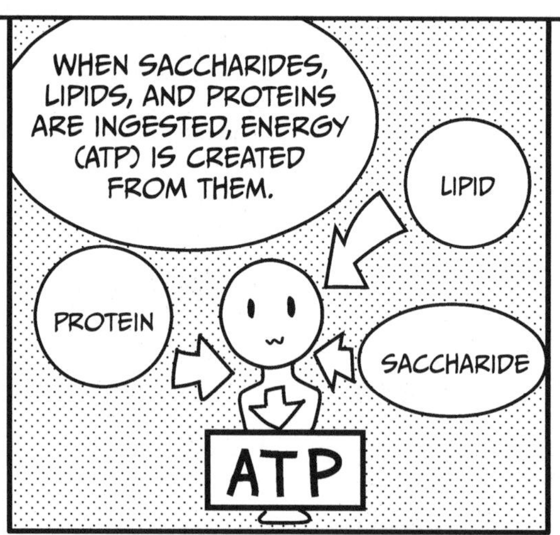

* MORE ACCURATELY, SINCE SOME OF THESE PARTICLES OF FOOD ARE DIRECTLY EXCRETED AFTER THEY ARE EATEN, INGESTED ENERGY IS THE TOTAL ENERGY THAT IS CREATED FROM THE SACCHARIDES, LIPIDS, AND PROTEINS THAT WERE ACTUALLY ABSORBED BY THE BODY.

SO THE MORE YOU EAT, THE MORE INGESTED ENERGY INCREASES.
THAT'S TRUE! EVEN WHEN I'M SLEEPING AT NIGHT OR JUST WATCHING TV...
ON THE OTHER HAND, AS LONG AS OUR BODIES ARE ALIVE, WE ARE ALWAYS USING ENERGY.
...MY BODY IS STILL CONSUMING THE MINIMUM REQUIRED ENERGY. THIS CONSUMPTION IS CALLED MY *BASAL METABOLISM*, RIGHT?
THAT'S RIGHT! AND OF COURSE IF YOU EXERCISE, YOU EXPEND MUCH MORE ENERGY.
URNG
RAH!
RAH!
THE TOTAL ENERGY USED IS CALLED *EXPENDED ENERGY*.

INGESTED ENERGY
EXPENDED ENERGY
ENERGY STORED AS FAT
IF EXPENDED ENERGY IS LESS THAN INGESTED ENERGY...
THAT LEFTOVER ENERGY IS STORED IN THE BODY IN THE FORM OF FAT.
HUFF
WHEW!
PUFF
I'VE GOT TO KEEP ACTIVE IF I WANT TO KEEP THOSE POUNDS OFF.
WHEEZE

ANIMALS PRESERVE FAT

Animals have a mechanism for preserving a fixed level of fat in the body. If they overeat and fat accumulates, that triggers a signal that tells the brain to reduce the amount that is eaten. Conversely, if the amout of fat is reduced, causing the animal to get hungry, it will eat until the fat level returns to its original amount.

Interesting! I guess that instinct must come from animals living in the wild, where life is severe. If it gets too fat, it won't be able to catch food, and if it gets too skinny, it could starve and die.

That's right. For example, the protein hormone *insulin* causes glucose in the blood to be absorbed into muscle or adipose tissue and stored as glycogen or fat. The result is that it lowers the blood sugar level. (Insulin is also used to treat diabetes.)

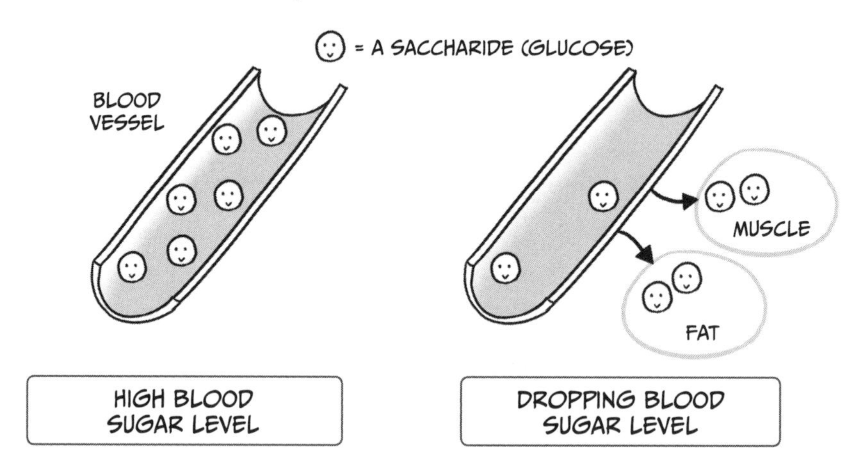

Oh, so "blood sugar level" is just the concentration of glucose in the blood! And if that glucose is absorbed into muscle or fat, the concentration in the blood will drop.

Also, the cell membranes of nerve cells in the hypothalamus (a special part of the brain) contain a protein called an *insulin receptor.**

* Insulin receptors actually exist in the cell membranes of many types of cells in the body, not just nerve cells.

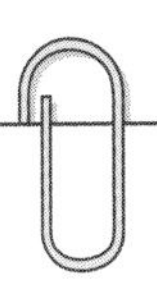

So insulin regulates the eating behavior and fat level of an animal via the hypothalamus.

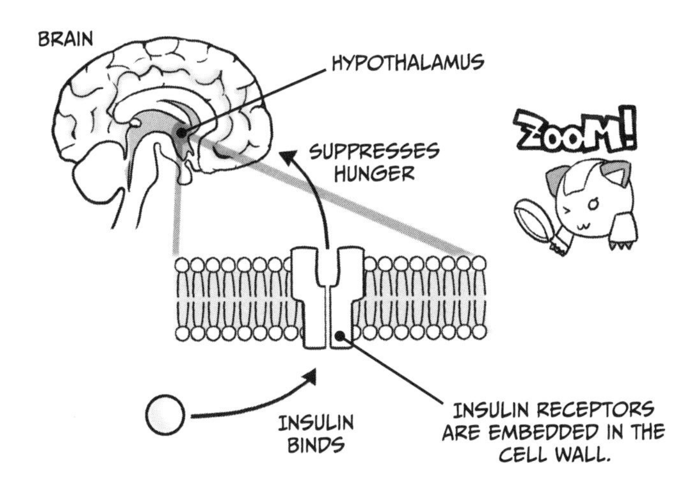

The urge to eat is suppressed by the nervous system when insulin binds with the insulin receptors in the hypothalamus.

The experiments scientists have done to prove this may surprise you—for instance, lab mice that were prevented from producing insulin receptors ended up becoming really obese.

Wow! Instructions go to my brain to keep me from eating too much? That's crazy—it's like my own body is playing a trick on me!

There's also a protein called *leptin*, which is only created in adipose tissue, that can notify the brain of an accumulation of fat via leptin receptors in the hypothalamus, much like insulin receptors. This also suppresses hunger.

I see. So both insulin and leptin are important proteins for controlling how much we want to eat.

That's right. And scientists have done experiments with leptin too—mice with deformed leptin genes end up just as tubby as the mice without insulin receptors.

That's awful! I wonder why scientists are always picking on mice, anyway...

If we compare the leptin concentration in the blood of nonobese people versus obese people, they both have a value that is proportional to the amount of fat they carry.

The amount of leptin created by people who overeat is no different than that of people who eat normally. Although the creation of leptin is linked with the suppression of eating in nonobese people, this is not the case in obese people. In other words, obese people are thought to have a resistance to leptin.

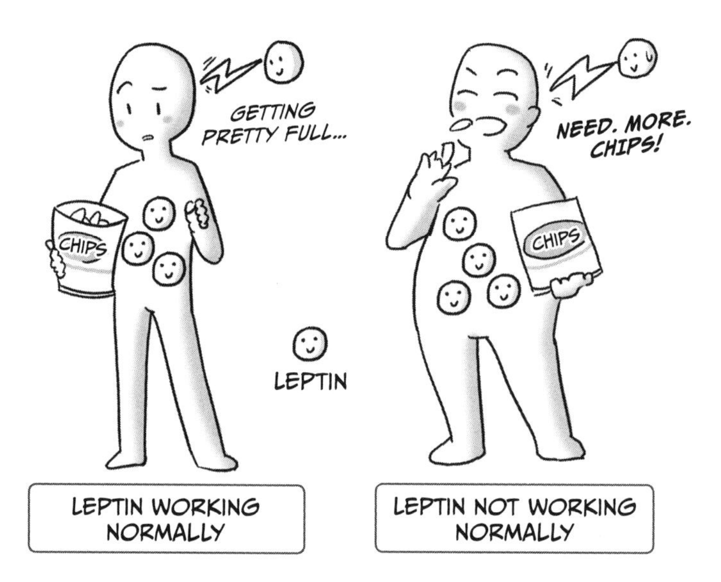

What a nightmare! If our insulin or leptin is ineffective, we'll eat and eat, and our appetite will never be satisfied...

There are also a number of other substances that help balance our fat levels and appetite, which we'll talk about next!

EXCESS SACCHARIDES BECOME FAT!
SO OBESITY IS THE RESULT OF A SIGNIFICANT BREAKDOWN OF THE METABOLIC HOMEOSTATIC* BALANCE.
IT'S A RESULT OF FAT BEING EXCESSIVELY ACCUMULATED.
IN SHORT...
YOU'VE EATEN TOO MUCH AND STORED TOO MUCH FAT.
よろ よろ
* HOMEOSTASIS IS THE MECHANISM IN WHICH SUBSTANCES ARE CONTINUALLY EXCHANGED WITH THE OUTSIDE WORLD TO KEEP AN INTERNAL ENVIRONMENT UNCHANGED.
NOW LET'S STUDY THE BIOCHEMICAL PROCESSES THAT ARE INVOLVED...
わーい♥
?
WHAT HAPPENED?!
...WHEN FAT ACCUMULATES!
THERE ARE TWO MAIN WAYS THAT FAT IS ACCUMULATED.
LIPID
SACCHARIDE
(1) WHEN INGESTED LIPIDS ARE DIRECTLY ACCUMULATED AS FAT...
AND (2) WHEN SACCHARIDES ARE CONVERTED TO FAT.
LIPID
SACCHARIDE
EW

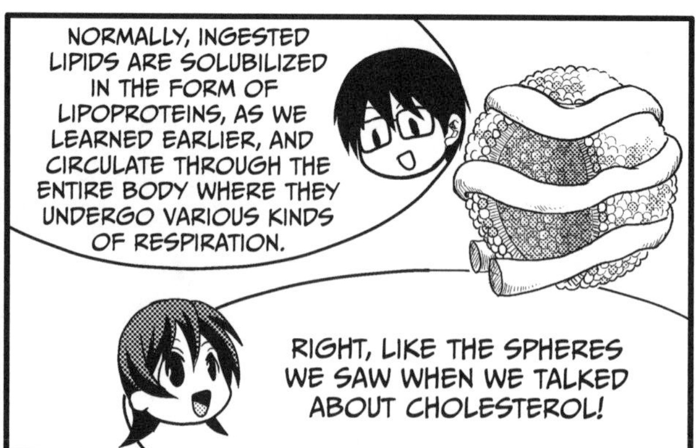

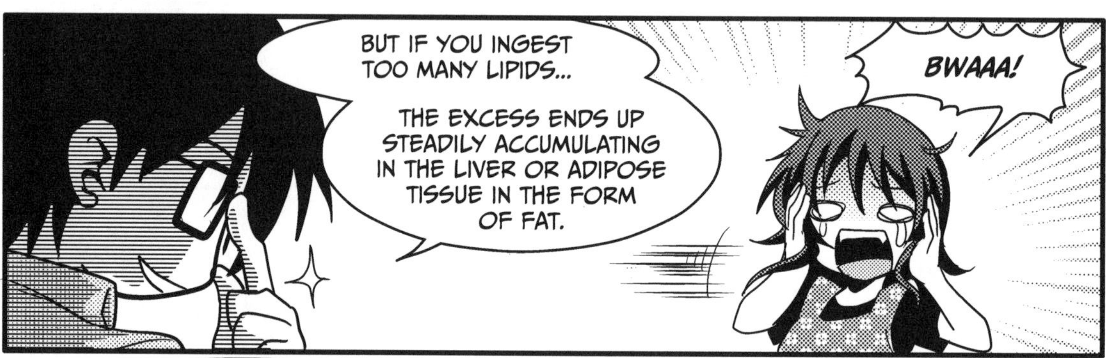

* HYDROLYSIS IS THE DECOMPOSITION OF A SUBSTANCE BY AN ENZYME USING WATER. FOR DETAILS, SEE PAGE 172.

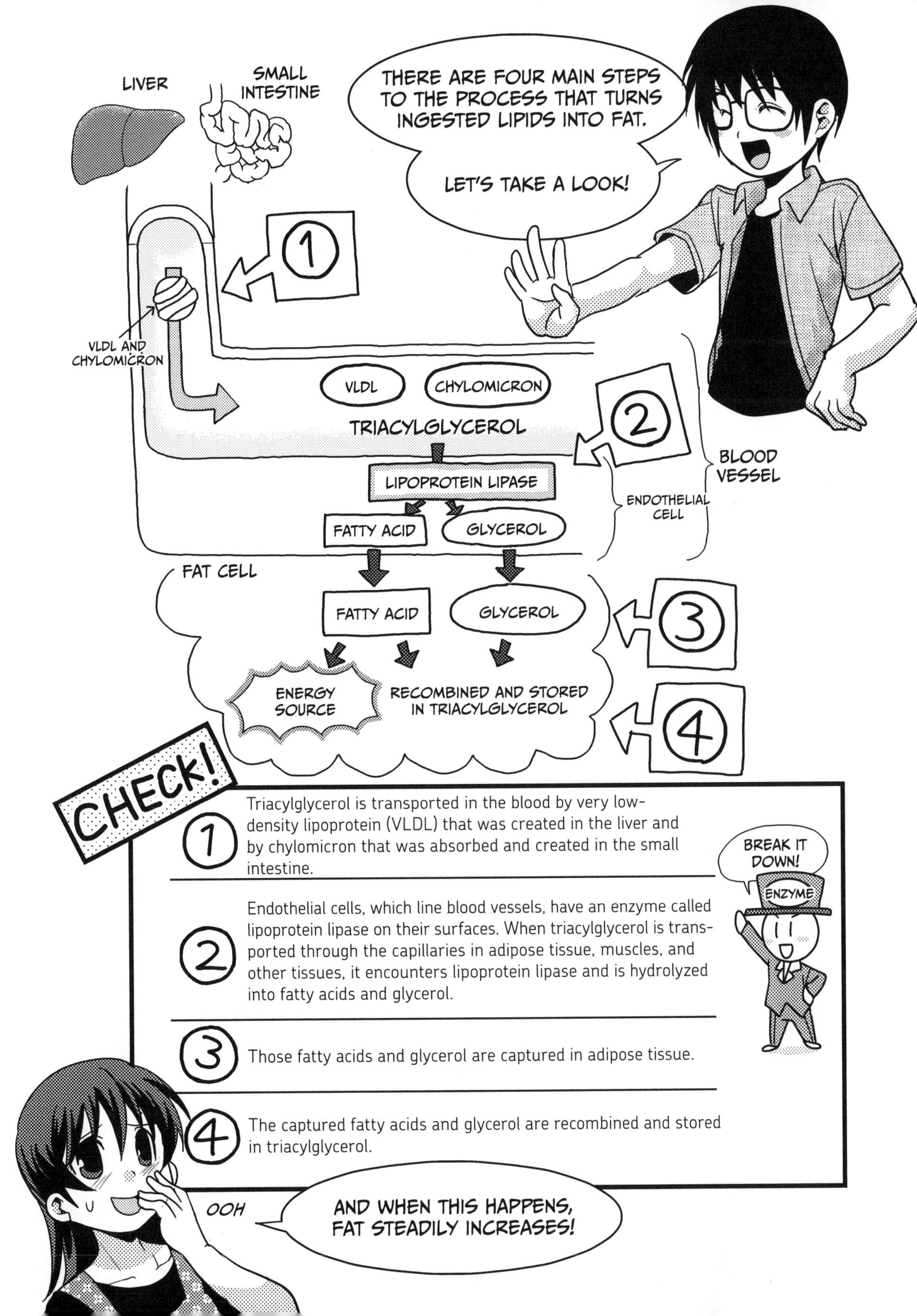
LIVER
SMALL INTESTINE
THERE ARE FOUR MAIN STEPS TO THE PROCESS THAT TURNS INGESTED LIPIDS INTO FAT.
LET'S TAKE A LOOK!
1
VLDL AND CHYLOMICRON
VLDL
CHYLOMICRON
TRIACYLGLYCEROL
2
LIPOPROTEIN LIPASE
BLOOD VESSEL
ENDOTHELIAL CELL
FATTY ACID
GLYCEROL
FAT CELL
FATTY ACID
GLYCEROL
3
ENERGY SOURCE
RECOMBINED AND STORED IN TRIACYLGLYCEROL
4
CHECK!
1 Triacylglycerol is transported in the blood by very low-density lipoprotein (VLDL) that was created in the liver and by chylomicron that was absorbed and created in the small intestine.
2 Endothelial cells, which line blood vessels, have an enzyme called lipoprotein lipase on their surfaces. When triacylglycerol is transported through the capillaries in adipose tissue, muscles, and other tissues, it encounters lipoprotein lipase and is hydrolyzed into fatty acids and glycerol.
3 Those fatty acids and glycerol are captured in adipose tissue.
4 The captured fatty acids and glycerol are recombined and stored in triacylglycerol.
BREAK IT DOWN!
ENZYME
OOH
AND WHEN THIS HAPPENS, FAT STEADILY INCREASES!

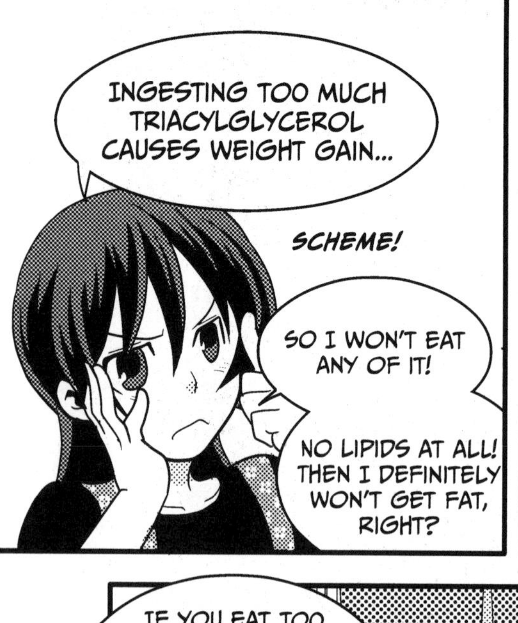

IF YOU EAT TOO MUCH RAMEN OR SPAGHETTI, YOU'LL BE INGESTING TOO MANY SACCHARIDES, AND YOU'LL GAIN WEIGHT.

THERE'S REALLY NO SHORTCUT HERE. A LITTLE BALANCE AND MODERATION IS ALL YOU NEED!

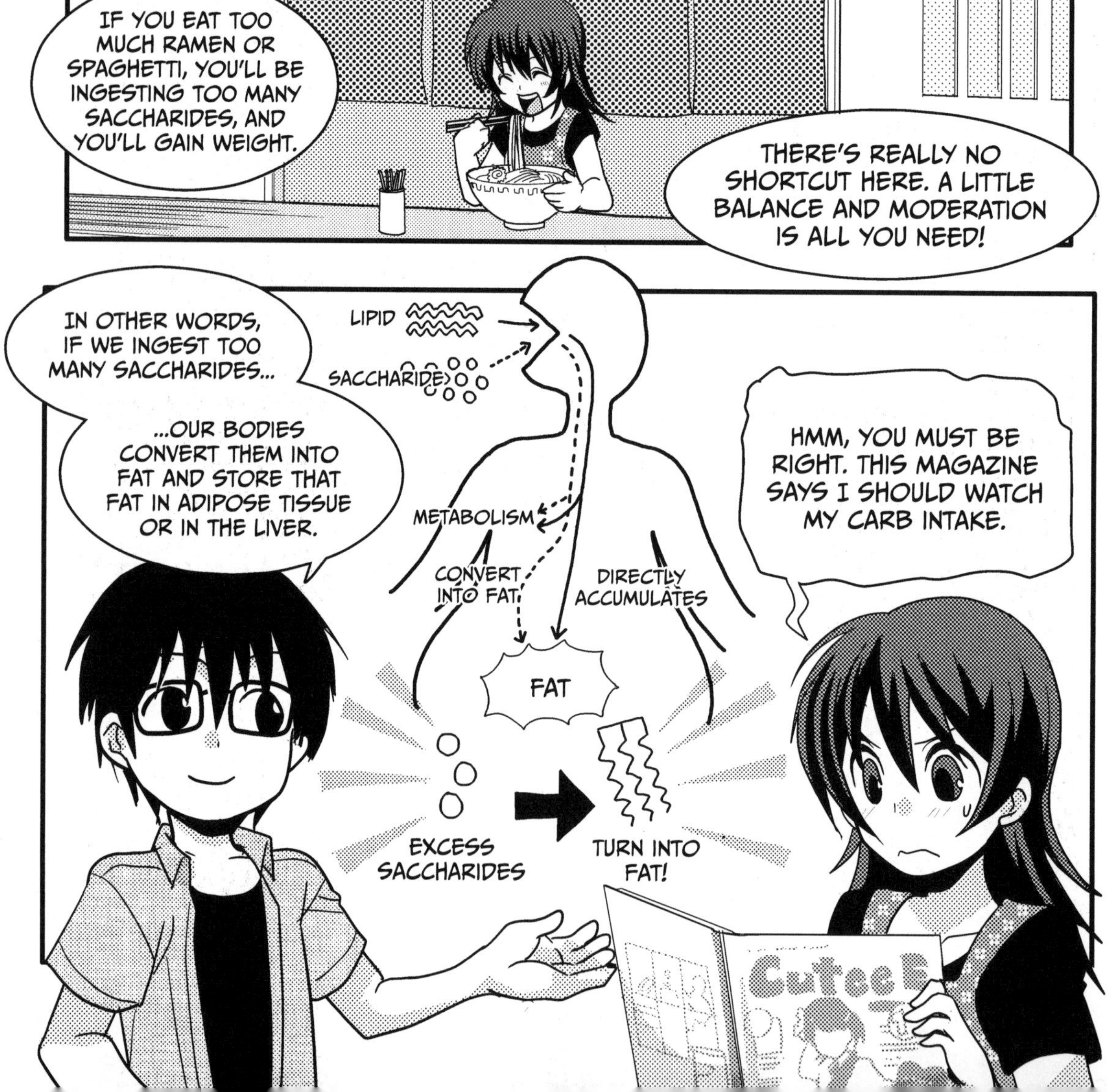

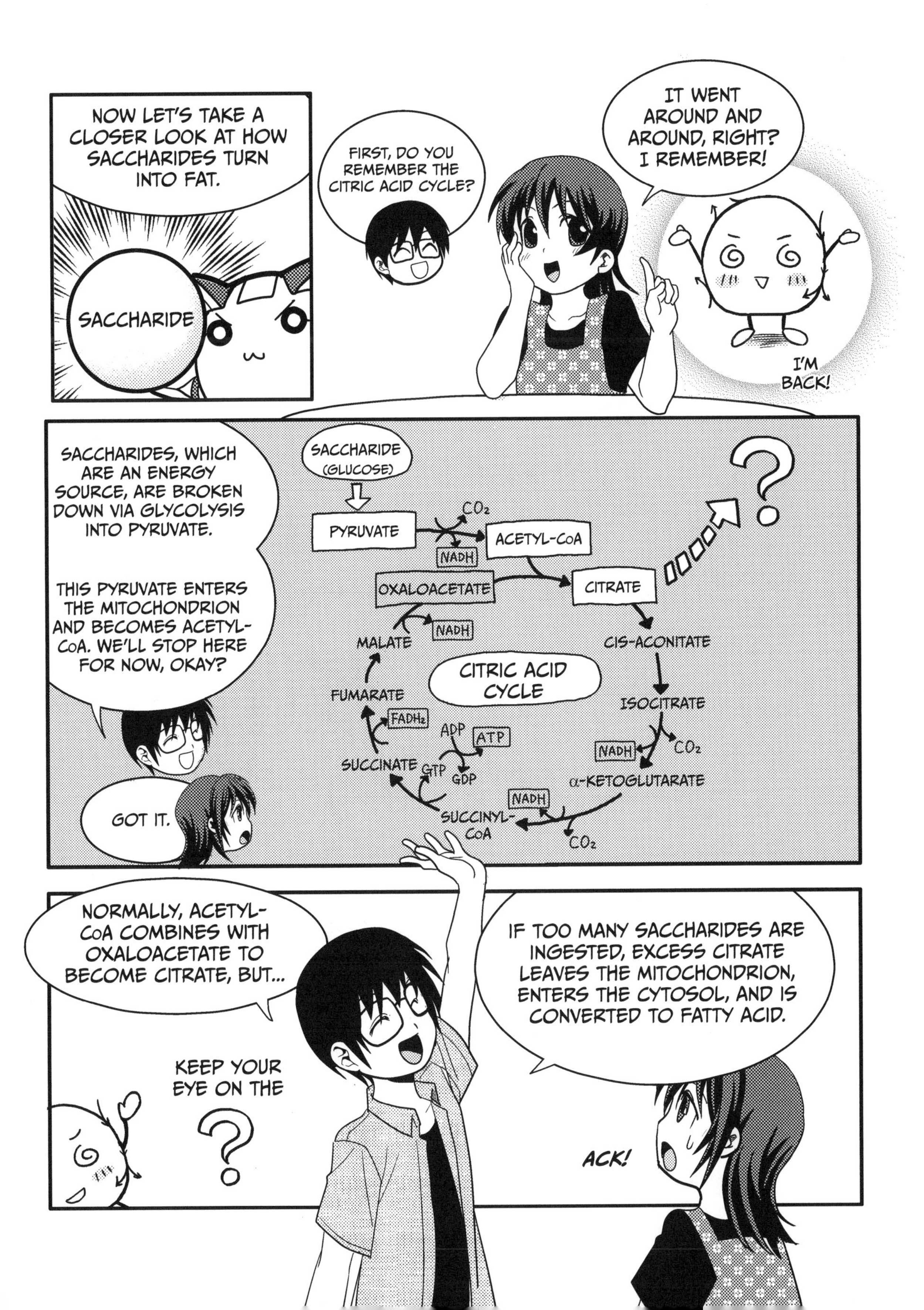
NOW LET'S TAKE A CLOSER LOOK AT HOW SACCHARIDES TURN INTO FAT.
SACCHARIDE
FIRST, DO YOU REMEMBER THE CITRIC ACID CYCLE?
IT WENT AROUND AND AROUND, RIGHT? I REMEMBER!
I'M BACK!
SACCHARIDES, WHICH ARE AN ENERGY SOURCE, ARE BROKEN DOWN VIA GLYCOLYSIS INTO PYRUVATE.
THIS PYRUVATE ENTERS THE MITOCHONDRION AND BECOMES ACETYL-CoA. WE'LL STOP HERE FOR NOW, OKAY?
GOT IT.
SACCHARIDE (GLUCOSE)
PYRUVATE
CO_2
NADH
ACETYL-CoA
OXALOACETATE
CITRATE
CIS-ACONITATE
ISOCITRATE
NADH
CO_2
α-KETOGLUTARATE
NADH
CO_2
SUCCINYL-CoA
GTP
GDP
ADP
ATP
SUCCINATE
$FADH_2$
FUMARATE
MALATE
NADH
CITRIC ACID CYCLE
?
NORMALLY, ACETYL-CoA COMBINES WITH OXALOACETATE TO BECOME CITRATE, BUT...
IF TOO MANY SACCHARIDES ARE INGESTED, EXCESS CITRATE LEAVES THE MITOCHONDRION, ENTERS THE CYTOSOL, AND IS CONVERTED TO FATTY ACID.
KEEP YOUR EYE ON THE
?
ACK!

WHEN TOO MANY SACCHARIDES ARE INGESTED, CITRATE EXITS INTO THE CYTOPLASM, IS BROKEN BACK DOWN INTO ACETYL-CoA, AND BECOMES THE SUBSTANCE CALLED *MALONYL-CoA*.
GLUCOSE
PYRUVATE
CYTOPLASM
ACETYL-CoA
CITRATE
OXALOACETATE
MITOCHONDRIA
PALMITIC ACID
MALONYL-CoA
ACETYL-CoA
ZOOM!
SINCE GLYCOLYSIS IS STILL PROCESSING SACCHARIDE, THE MALONYL-CoA, ACETYL-CoA, AND PALMITIC ACID WILL ACCUMULATE, WON'T THEY?
OOOOH
CITRATE DOES NOT ENTER THE CITRIC ACID CYCLE!
THAT'S RIGHT, AND PALMITIC ACID IS A FATTY ACID!
IF MALONYL-CoA IS PRODUCED, IT ENDS UP BEING DIRECTLY CONVERTED INTO A FATTY ACID.
EEP!
YOU SHOULD ALSO BE CAREFUL ABOUT HOW MUCH TABLE SUGAR (FRUCTOSE) YOU EAT.
SUGAR
FRUCTOSE PROMOTES FATTY ACID SYNTHESIS IN THE LIVER AND IS ONE OF THE CONTRIBUTING FACTORS TO OBESITY.
WOE IS ME... BETRAYED BY SWEET, SWEET SUGAR...

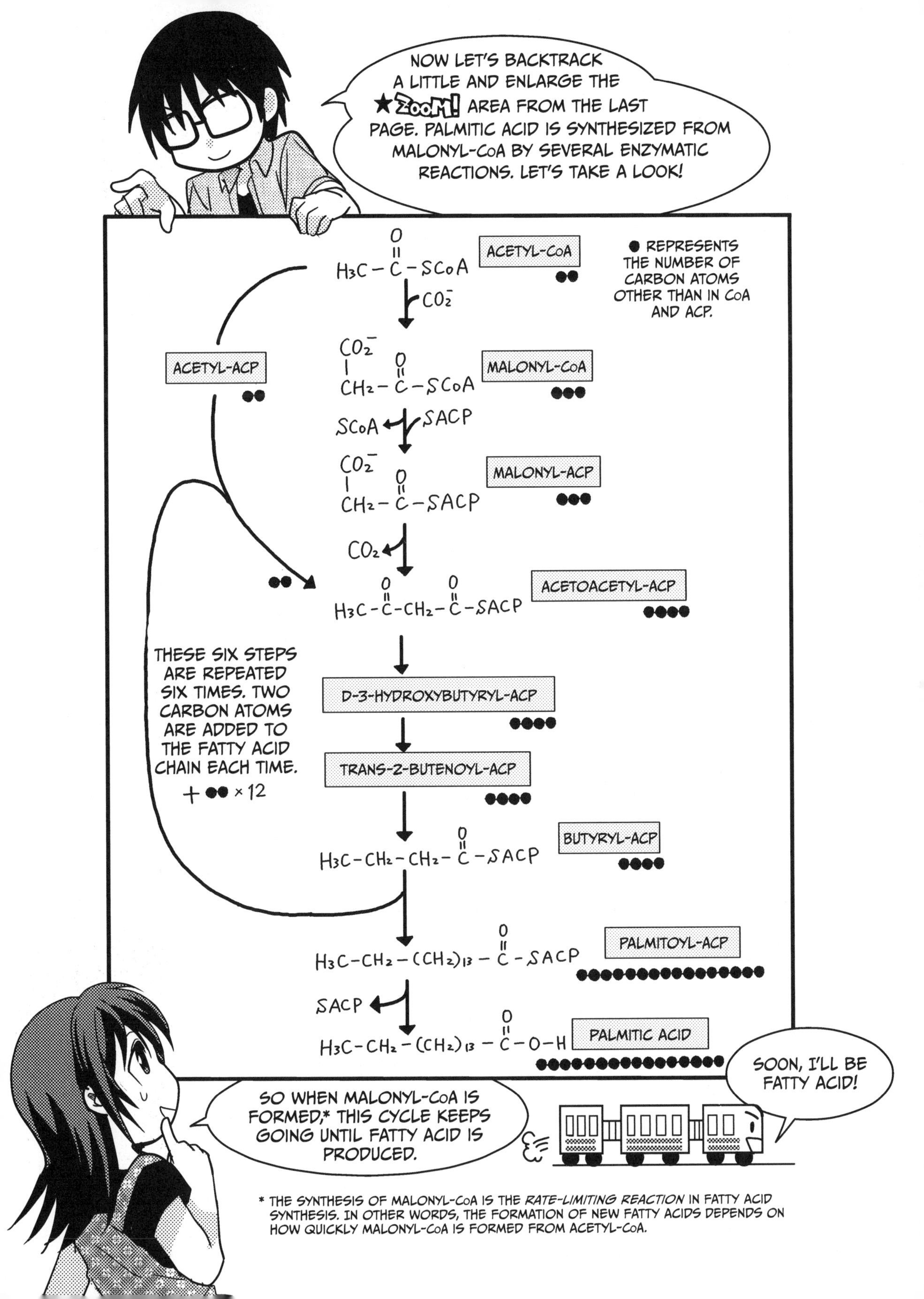

* THE SYNTHESIS OF MALONYL-CoA IS THE *RATE-LIMITING REACTION* IN FATTY ACID SYNTHESIS. IN OTHER WORDS, THE FORMATION OF NEW FATTY ACIDS DEPENDS ON HOW QUICKLY MALONYL-CoA IS FORMED FROM ACETYL-CoA.

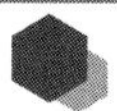

WHEN FAT IS USED AS AN ENERGY SOURCE

Okay, now I understand how fat is produced, but what I really want to know is how to **get rid of** the fat I'm producing.

To lose weight, you just have to steadily use up your stored fat.

To do this, you need to remember one important thing: **When both saccharides and lipids are present, the saccharides will be used as an energy source first**.

If you eat foods containing lots of saccharides and lipids, your blood sugar level will rise. The saccharides are used first for energy production, and the lipids are stored in adipose tissue.

However, when the saccharides are used up and your blood sugar level decreases, the stored lipids will gradually start getting metabolized. So when your hungry stomach growls, lipids are actively being consumed.

AFTER SACCHARIDES ARE USED UP...

PUFF
PUFF

SACCHARIDE
BLOOD VESSEL

LIPIDS ARE USED.

SO HUNGRY...

NOW IT'S OUR TURN!
FAT

Hmm, I guess that means I've got to exercise for quite a while before those pounds start to burn off.

Now let's talk about how fat is metabolized.

First, the fat in adipose tissue, **triacylglycerol**, is broken down into **fatty acid** and **glycerol** by an enzyme called *hydrolase*. Hydrolase is a hormone-sensitive lipase—note that this is different from a lipoprotein lipase, which we talked about on page 112.

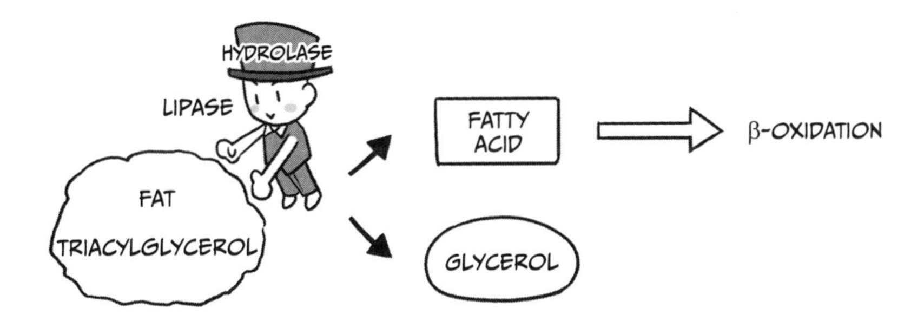

Fatty acid is released into the blood and transported to the various organs and muscles of the body, where it undergoes a chemical reaction called β-*oxidation* (that's *beta* oxidation).

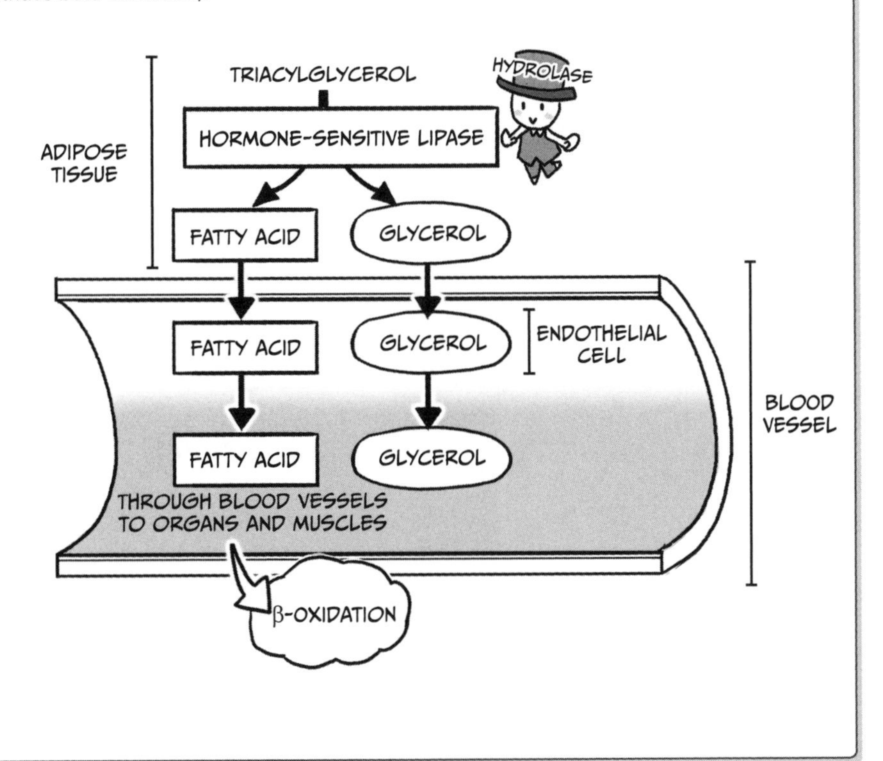

Um...what in the world is β-oxidation?

Relax, I'm getting to it! Take a look at this diagram.

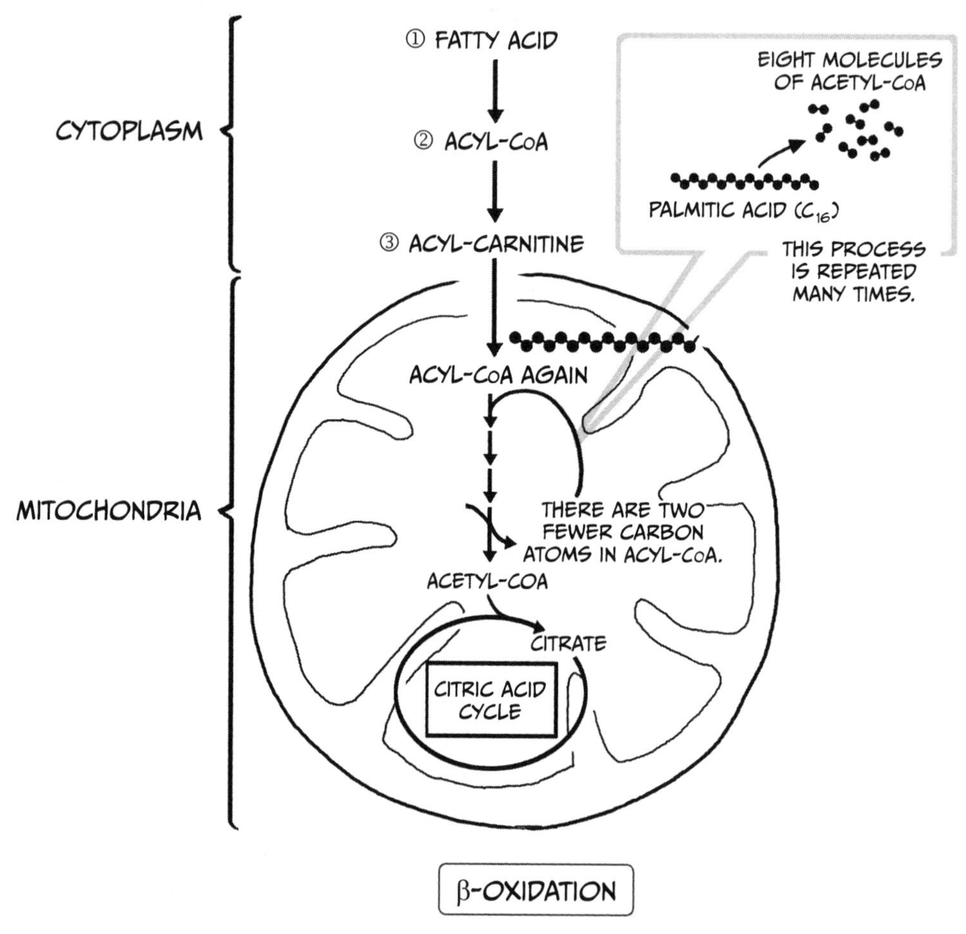

β-OXIDATION

Fatty acid becomes acetyl-CoA through β-oxidation. Remember when we talked about acetyl-CoA earlier? (See page 115.)

It's the stuff that appears in the first step of the citric acid cycle, right?

You got it!

First, the fatty acid is *activated* by the attachment of CoA. This fatty acid–CoA compound is called *acyl-CoA*. A carnitine is then added to form *acyl-carnitine*, which is shuttled into the cell. It then enters the mitochondrion and is broken down into acetyl-CoA.

Fatty acids are long carbon chains (10+), but acetyl-CoA has only two carbon atoms.

The fatty acid is broken down by β-oxidation so that one molecule of acetyl-CoA (two carbon atoms) is detached each time. The process is called β-oxidation because the CoA is attached to the second-to-last carbon atom of the fatty acid (the β carbon).

Ultimately, all carbon atoms in the fatty acid will become acetyl-CoA. The following figure shows the β-oxidation of palmitic acid, which has 16 carbon atoms.

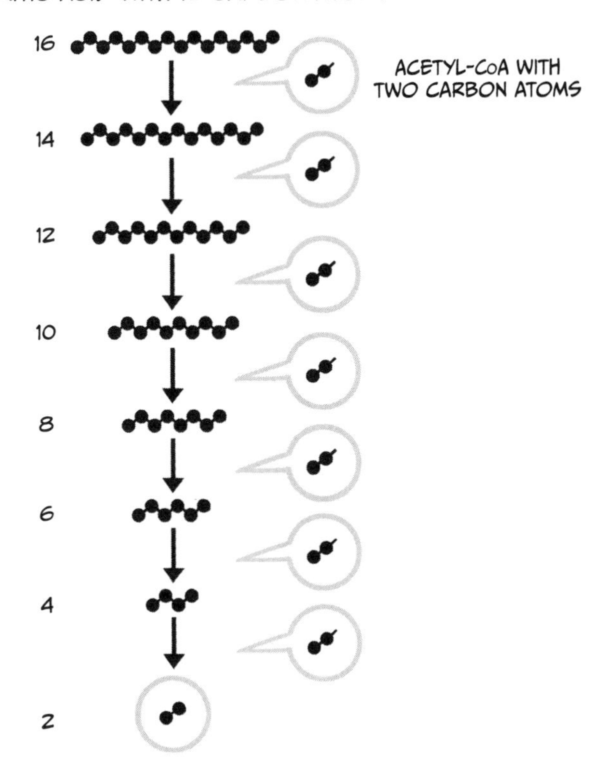

The β-oxidation cycle is repeated seven times to make eight molecules of acetyl-CoA!

Right, and since they're already in the mitochondrion, they can directly enter the citric acid cycle there to create ATP.

So when you're burning fat and dieting, this crazy process is breaking down fatty acids inside your body. Wow!

Earlier, I told you how palmitic acid is produced from malonyl-CoA during fatty acid synthesis. When this palmitic acid is broken down, a total of 129 molecules of ATP are produced through the TCA cycle and electron transport chain.

Wait...if I remember correctly, weren't 38 molecules of ATP the most that could be created from 1 molecule of glucose? 129 is an awful lot, isn't it?

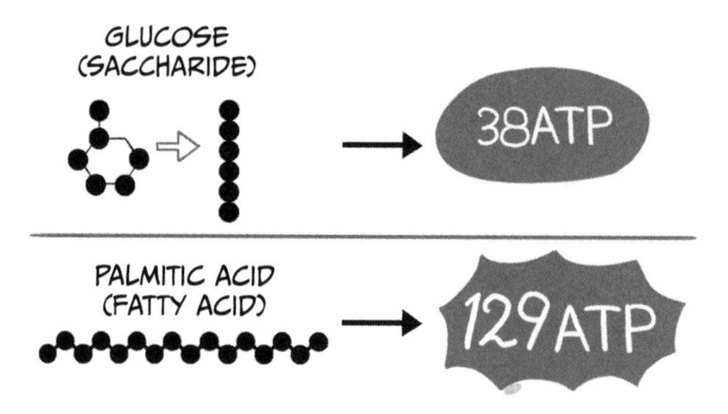

Yup, that's true. Fatty acid is a very efficient storage material.

Hmmm. That must be why it takes so much energy to burn it away! Biochemically speaking, dieting is a lot of work...

MYSTERY 2

WHY DO YOU GAIN WEIGHT IF YOU OVEREAT?

- When expended energy is less than ingested energy, the body stores the excess ingested energy as fat.
- There are two main ways that fat is formed. One is the process in which ingested lipids (that is, triacylglycerol) are directly accumulated as fat. The other converts saccharides into fat.
- Fat is an efficient energy storage material.

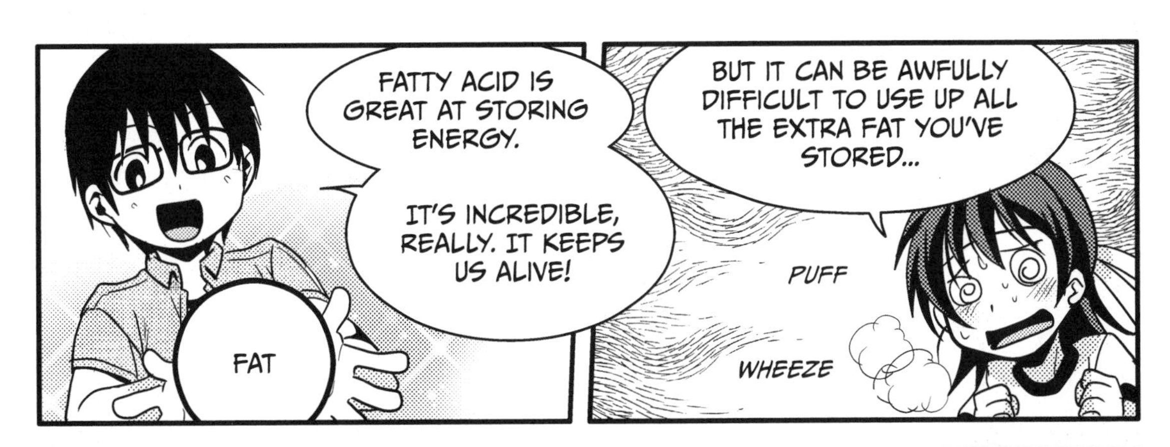

3. What Is Blood Type?

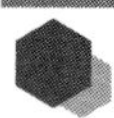

BLOOD TYPE

The third mystery is: **What is blood type?** This should be fun!

Even if we totally avoid the science of blood types, they can still be an interesting way to classify people. For example, in Japan it's common to tell someone's fortune based on their blood type.

That's right! A Type A person is often said to be calm, composed, and serious.

Incidentally, I am an unconventional, do-it-my-own-way Type B!

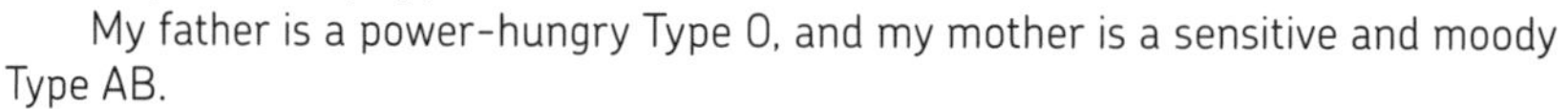

My father is a power-hungry Type O, and my mother is a sensitive and moody Type AB.

Hmm, that makes sense. Children of parents with Type O and Type AB should be either Type A or Type B but not Type O or Type AB. So if you were to have a sibling, he or she would be Type A or Type B.

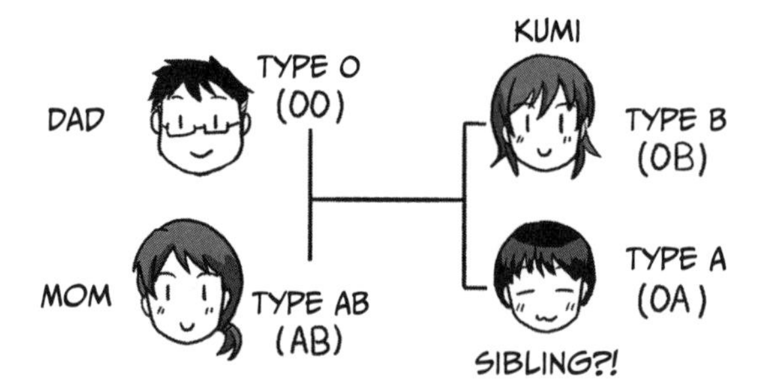

If I had a younger brother who was Type A, my family would have people of all four blood types.

But it's a little strange, isn't it? Members of one family can have different blood types, but people from completely different families can have the same blood type!

What exactly **is** blood type, anyway?

HOW IS BLOOD TYPE DETERMINED?

Kumi, you've probably learned about red blood cells in school, right?

Yeah! They're the cells that give blood its color. They look something like this:

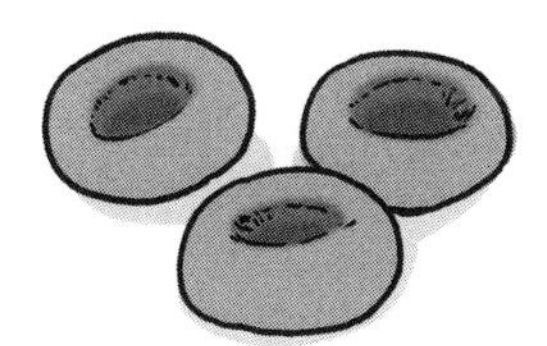

You got it. Blood type is determined by the type of saccharide molecules that protrude from the surface of red blood cells. The surface of many cells, including red blood cells, is covered by a *sugar coating* (*glycocalyx*) made up of saccharides.

Jeez, we're back to saccharides again? Though I do like the sound of this "sugar coating"...

Remember the lipid bilayer of the cell membrane? Proteins are embedded in various spots in that bilayer, and saccharide molecules are often attached to their outer surfaces. Collectively, these saccharide molecules make up the cell's sugar coating.

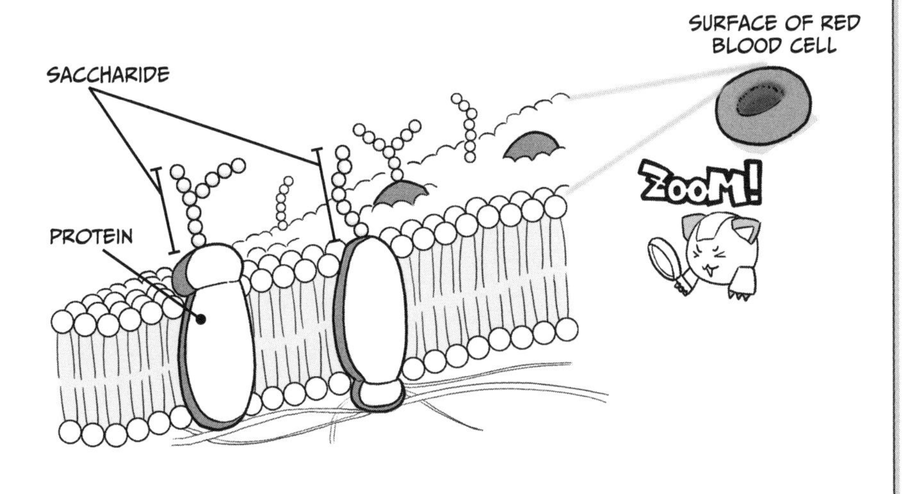

How strange! They look almost like little sea anemones or something, reaching out from the bottom of the ocean. And they're all over our blood cells?

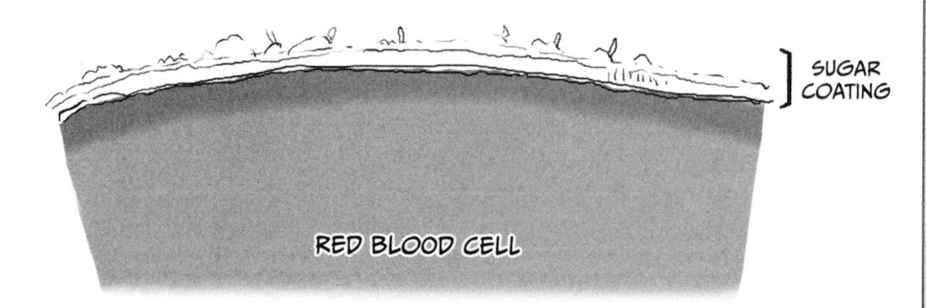

That's right. There are actually over 100 different antigens that can coat red blood cells and be used to classify blood types, so many different *blood group systems* are possible. The most famous is the *ABO blood group system*, which was discovered in 1900 by the Austrian immunologist Karl Landsteiner.

That's the system that we use now, right? With Type A, Type B, Type AB, and Type O?

Right again! The ABO blood group system is based on **three types of saccharide molecules** present on the surface of red blood cells. Each of these has a structure called a *sugar chain*, which consists of several monosaccharides connected together.

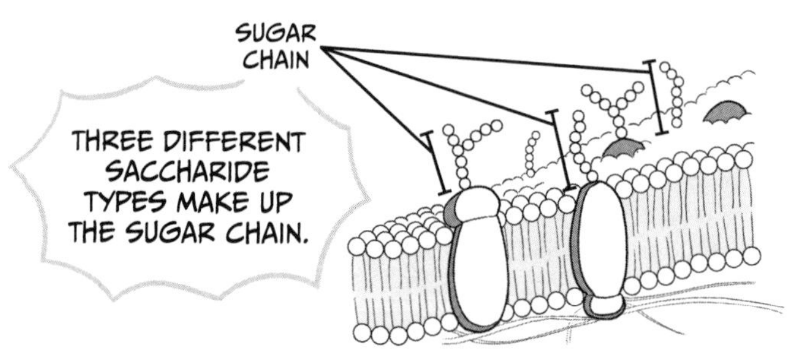

Three types of "anemones" waving around down there, huh?

If we look at the left side of the diagrams below, we can see that the tip for each type of sugar chain is different.

For the sugar chain of Type A blood, the tip is *GalNAc*.

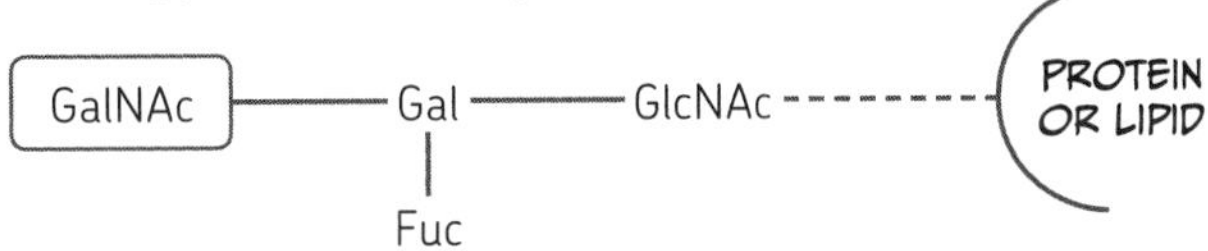

For the sugar chain of Type B blood, the tip is *Gal*.

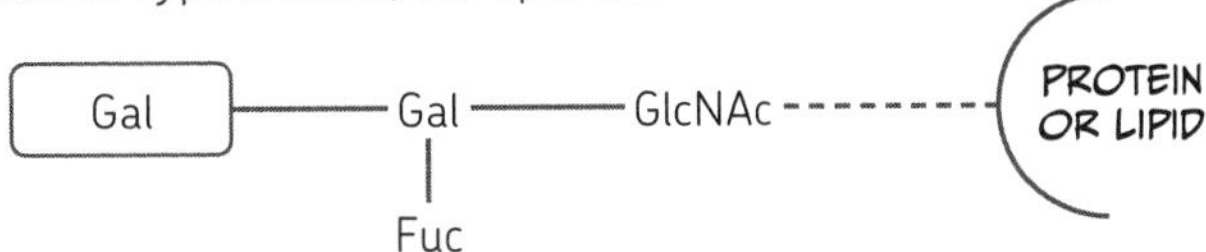

For the sugar chain of Type O blood, there is no tip!

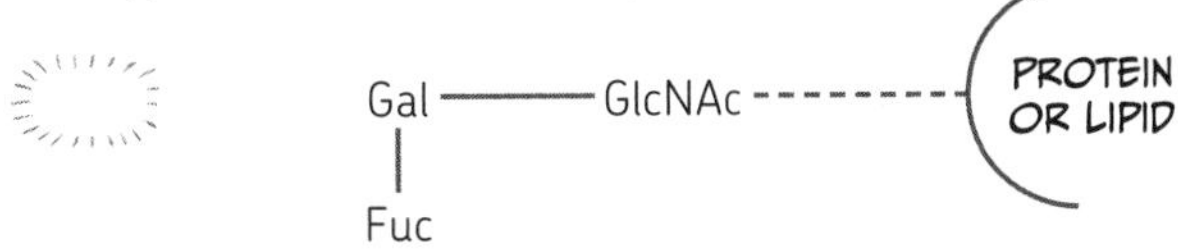

NAMES OF SACCHARIDES

GalNAc	: *N*-acetylgalactosamine
Gal	: Galactose
Fuc	: Fucose
GlcNAc	: *N*-acetylglucosamine

Oooh, so that's where those three types come from!

That's right! And people with Type AB blood have sugar chains of both Type A and Type B.

Wow, that's all there is to it? Differences in sugar chains? How weird...

But wait...why does one person become Type A while another becomes Type B? What decides which blood type a person will have?

Well, blood type is determined by a certain gene, and an enzyme is created by that gene! (We'll come back to this later, but you can skip ahead to page 169 for more details, if you'd like.)

Again with the enzymes! You mentioned them when we discussed fat, and now again when we're talking about blood types? They must be really important.

Definitely. We'll have to get the professor to tell us more about enzymes when we see her again later.

Yeah! By the way, is it possible that the differences in these sugar chains could actually be related to a person's personality? Can you really tell a person's fortune from their blood type?

Hmm, good question. I guess if you went out on a limb, you could say that the gene that determines blood type might also have some kind of influence on nerve cells. But claiming that that might actually influence one's personality is a pretty big leap of logic, and there's no scientific evidence to support it.

That's a relief. I'd hate to think that my entire personality is controlled by a bunch of dinky little sugar chains.

I don't care what fortune-tellers say—there's no one in the world with a personality like yours, Kumi...

WHAT IS BLOOD TYPE?

- The four blood types (A, B, AB, and O) are based on the ABO blood group system.
- The ABO blood group system classifies types according to the differences between certain sugar chains on the surface of red blood cells.
- So far, no real evidence has been found to support the idea that the differences in sugar chains affect personality, so don't let those fortune-tellers boss you around!

* FORTUNE-TELLING BY BLOOD TYPE

4. Why Does Fruit Get Sweeter as It Ripens?

WHAT TYPES OF SUGAR ARE IN FRUIT?

Time for the fourth mystery: **Why does fruit get sweeter as it ripens?**

Speaking of fruit, we just got some fresh pears at my house...

Mmmm, they are ripe, sweet, and delicious.

Actually, that's true for many different kinds of fruit. Mandarin oranges, grapes, cantaloupes, and watermelons all get tastier as they ripen.

You've got that right. When mandarin oranges are still a little greenish, they can be way too acidic, but ripe ones are really sweet. And the melon that you brought over earlier was in season and perfectly ripe!

Yep, but what does "ripe" mean, biochemically speaking? We know that ripe fruit is sweeter, but why?

The reason is that three types of sugars are contained in large quantities in fruits and berries: sucrose (table sugar), fructose (fruit sugar), and glucose (grape sugar*).

Weird! I would have expected them to have only fruit sugar. We are talking about fruit, after all...

* Despite the name, "grape" sugar is found in many fruits and berries, not just grapes.

MONOSACCHARIDES, OLIGOSACCHARIDES, AND POLYSACCHARIDES

Earlier we said that sucrose, glucose, and fructose all have different structures. (See page 63 for details.)

Right! There are a bunch of different kinds of saccharides, and I love eating all of them.

Well, now it's time to learn more about those different saccharides!

The basic unit of saccharides is called a *monosaccharide*, which is formed by at least three carbon atoms connected together.

Glucose and fructose are monosaccharides that consist of six carbon atoms. If two or more monosaccharides are attached, they become an *oligosaccharide*.

Although sucrose is an oligosaccharide, since it's formed by connecting only two monosaccharides, it's called a *disaccharide*. Here's what these saccharides look like:

GLUCOSE

FRUCTOSE

SUCROSE

Oh, so sucrose is made up of one glucose and one fructose connected together!

That's right! There are also saccharides made of many monosaccharides, which form extremely long molecules or complex branching structures. These are called *polysaccharides*.

Can you think of any polysaccharides that we might be familiar with?

Oh! We talked about potatoes and rice earlier, and they were chock-full of saccharides. So...starch is a polysaccharide, right?

That's right. Starch consists of many glucose monosaccharides connected together. Plants store glucose in this form after it's created during photosynthesis.

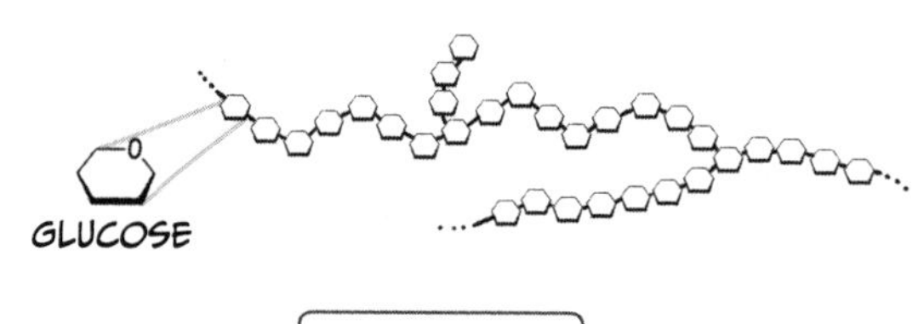

STARCH

Our bodies (and those of animals) also contain a "storage material" like starch. It's called glycogen, and it's produced mainly by the liver or muscles by connecting together excess glucose molecules to store them for later.

Oh yeah, I remember you mentioning glycogen earlier, but now it actually makes sense!

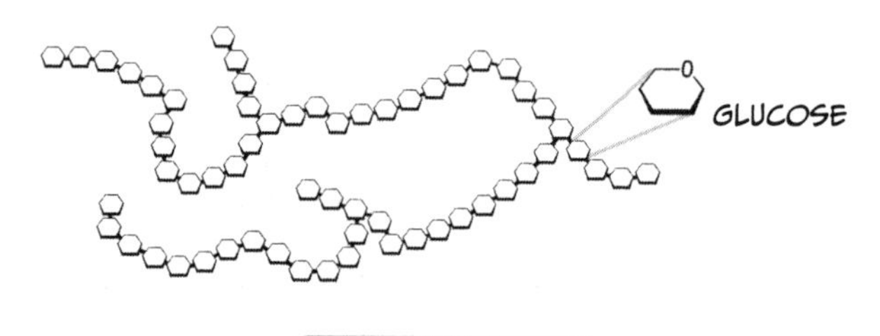

GLYCOGEN

Other types of polysaccharides include cellulose and chitin. Cellulose is the main component of plant cell walls, and chitin is the main component of the hard shells of crustaceans, like shrimp and crabs. Mushrooms also use chitin as a structural material.

HOW FRUITS BECOME SWEET

Now, let's get back to fruit. Fruits, such as mandarin oranges and melons, get sweeter and more delicious as they ripen. Why is that?

Hmm. When I was shopping for strawberries and mandarin oranges at the supermarket earlier, there was a sign that said, "Sugar Content: 11–12%."

I suppose if fruit gets sweeter as it ripens, that just means that the sugar content has increased, right? Biochemically speaking, we must be talking about a change in the saccharides.

You got it. So...let's talk about saccharides!

Take a look at the graphs below. Before they're ripe, citrus fruits, such as mandarin oranges, contain roughly an equal amount of glucose, fructose, and sucrose. But as they ripen, the relative amount of sucrose steadily increases. In a Japanese pear, all three types increase.

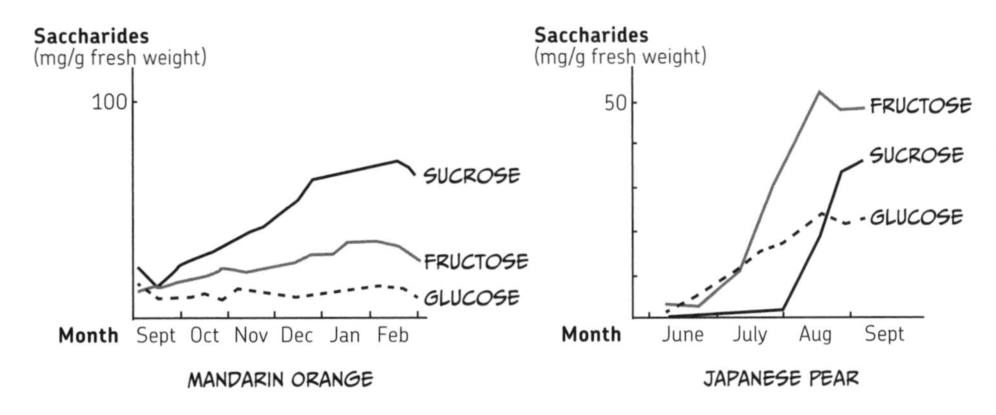

Source: Saburo Ito, Editor, *Science of Fruit*, Asakura Publishing Co., Ltd. (1991)

When fruit ripens, polysaccharides, like starch, are broken down into monosaccharides, like glucose, and the activity of *sucrose-phosphate synthase*, an enzyme that synthesizes sucrose in fruit, increases while the activity of *invertase*, which breaks down sucrose, decreases.

Oh...so sucrose-phosphate synthase is what makes sucrose by combining glucose and fructose .

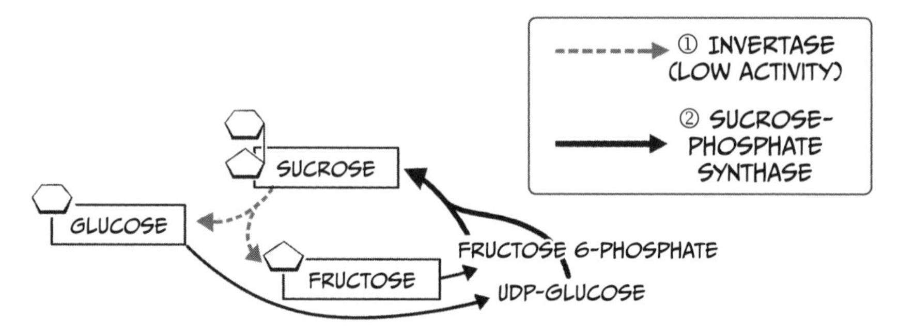

So does that mean that a fruit gets sweeter as more sucrose is produced?

Well, glucose, fructose, and sucrose are all sweet. Of the three, fructose is the sweetest, followed by sucrose, and then glucose.

DEGREES OF SWEETNESS, WHEN THE SWEETNESS OF GLUCOSE IS CONSIDERED TO BE 1.

Wow!

So as fructose or sucrose increases, the fruit gets sweeter and becomes ripe. For example, citrus fruits are best picked in winter, after their sucrose content increases and they're at their most sweet and delicious. For Japanese pears, on the other hand, the difference in the amounts of these sugars is more striking; as the fruit ripens, the fructose and sucrose both increase suddenly, as polysaccharides are broken down.

Similar to citrus fruits, the sweetness of melons depends mostly on sucrose content, and they'll be most delicious when sucrose levels are highest.

When you look at the graphs on the previous page, you can see that the fructose or sucrose increases, so the fruit gets really sweet in the harvesting season.

WHY DOES FRUIT BECOME SWEET?

- Fruit contain three types of sweet saccharides: sucrose, fructose, and glucose.
- As fruit ripens, enzymes become active, changing the amounts of the three saccharides.
- Fructose and sucrose are sweeter than glucose, so fruit becomes sweeter as the amount of fructose or sucrose increases!

* ADDITIONALLY, SUBSTANCES SUCH AS SORBITOL, XYLOSE, AND ORGANIC ACIDS ARE ALSO RELATED TO THE SWEETNESS OR ACIDITY OF FRUIT.

5. Why Are Mochi Rice Cakes Springy?

DIFFERENCES BETWEEN NORMAL RICE AND MOCHI RICE

The final mystery is: **Why are mochi rice cakes springy?**

I love mochi, so this is my kind of mystery.

First, let's talk about how mochi is made. Did you know that mochi is made from special rice, which is springier than normal rice?

Of course I know that! I've got serious mochi-making experience. I've even participated in a traditional "mochi-pounding" ceremony!

But what makes mochi rice springier than normal rice?

It's all because of the structural differences of the starches that make up the rice.

Rice is 75% starch, so slight variances in its starch composition can seriously affect its physical properties.

As you can see in the figure below, "normal," non-glutinous rice contains the two starches amylose and amylopectin. Amylose makes up about 17 to 22% of the starch, and the rest is amylopectin.

17–22% 100%

NON-GLUTINOUS RICE

MOCHI RICE

AMYLOSE

AMYLOPECTIN

COMPOSITION OF THE STARCH CONTAINED IN RICE

However, mochi rice starch only contains amylopectin.

Okay, the graph makes perfect sense, but what does that actually mean?

Just hang on, it should become clear in a second.

So...amylose and amylopectin are both made up of glucose molecules connected together to form polysaccharides.

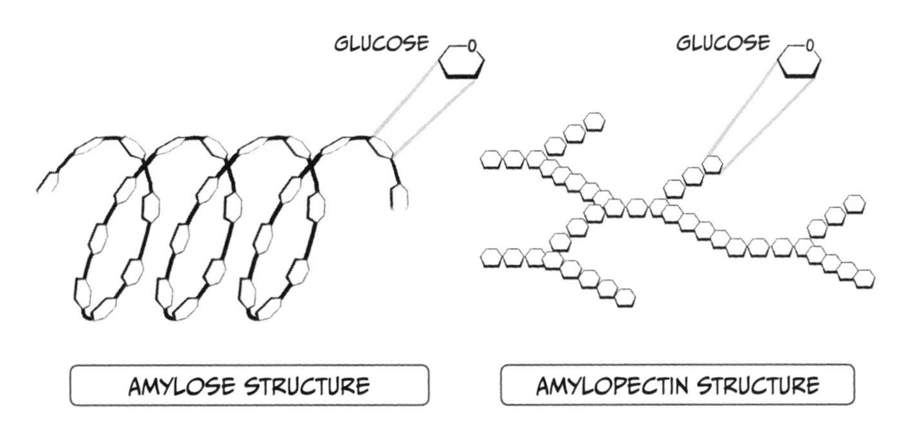

In other words, they're both made up of the same raw materials!

That's right. The only real difference is the way that their glucose molecules are connected together.

So the "mystery of springy mochi" is **actually** the "mystery of monosaccharides connecting in different ways to form polysaccharides and oligosaccharides." I guess the name's not quite as catchy, but it's still a very interesting subject.

Aha! So the secret of mochi's springiness is hidden in how the glucose connects together!

THE DIFFERENCE BETWEEN AMYLOSE AND AMYLOPECTIN

So we know that the difference between these two starches is the way in which their glucose molecules are connected. Specifically, the shape of amylose is **straight**, and the shape of amylopectin is **branched**.

Straight? Branched? I have no idea what you're talking about...

Amylose is formed when glucose molecules are connected in a straight line via a method called the α*(1→4) glycosidic bond*.

α(1→4) glycosidic bond

AMYLOSE

Oh, the glucose molecules are lined up in a row. That makes sense.

However, in amylopectin, there are places where glucose molecules are connected via another process called the α*(1→6) glycosidic bond*.

CH_2OH CH_2OH CH_2OH

α(1→6) glycosidic bond

CH_2OH CH_2OH CH_2 CH_2OH CH_2OH

AMYLOPECTIN

Oh, I see. That bond connects the molecules vertically! That must be what causes amylopectin to branch off in different directions.

You got it, Kumi. These two types of bonds are what give amylopectin its more complex, branched structure and amylose its straightforward, linked structure.

I totally understand! Man, I thought this stuff was supposed to be difficult.

Because of this branched structure, when amylopectin is viewed from a distance, it has a distinct "fringed" shape.

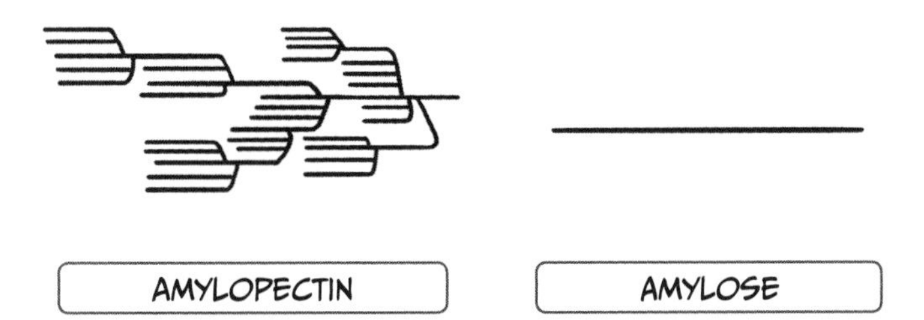

Mochi rice, which is mostly made up of amylopectin, becomes very viscous when cooked because the branches stick together. In other words, it feels very springy and stretchy.

So since this "fringed" kind of starch is more elastic than the straight kind, mochi made with this special rice is springy and delicious!

By the way, even in normal, non-glutinous rice, the stickiness varies according to the amount of amylose it contains.

WHAT DO THE NUMBERS MEAN IN $\alpha(1\rightarrow4)$ AND $\alpha(1\rightarrow6)$?

Since we've come this far, let's take this opportunity to learn what the numbers mean in the two types of bonds we discussed: $\alpha(1\rightarrow4)$ and $\alpha(1\rightarrow6)$.

Yeah, I was wondering about that, but I was kind of afraid to ask...

This may seem a little out of the blue, but...do you like baseball?

Um, I guess so. My dad is a big fan, so I see a lot of games on TV.

Numbering the bases on the field makes the game much easier to understand, right?

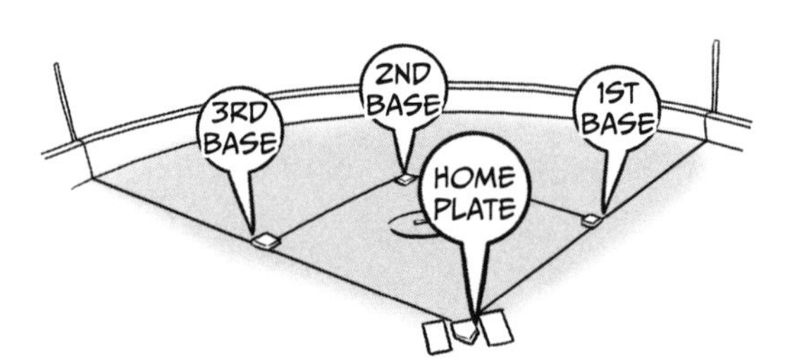

If they weren't numbered, it would make describing a game a lot trickier. For instance, we'd have to call third base "the base to the catcher's left." And I don't even want to **think** about how we'd talk about a triple play!

It would make the announcer's life difficult, that's for sure.

We learned that glucose and fructose have six carbon atoms each, right?

Numbers have been assigned to each of those atoms just like the bases in baseball!

Look carefully at the carbon atoms (C) in the following figure. This is the ring structure of glucose, and the numbers 1 through 6 have been assigned as shown.

⑥CH_2OH
⑤C — O
H H
④C C①
H OH H
HO ③C — C② OH
H OH

Wait, I think I get it!

The numbers in the α(1→4) glycosidic bond and the α(1→6) glycosidic bond refer to those numbers!

Right you are! The α(1→4) glycosidic bond means that the first carbon of one glucose is attached via a glycosidic bond to the fourth carbon of the neighboring glucose.*

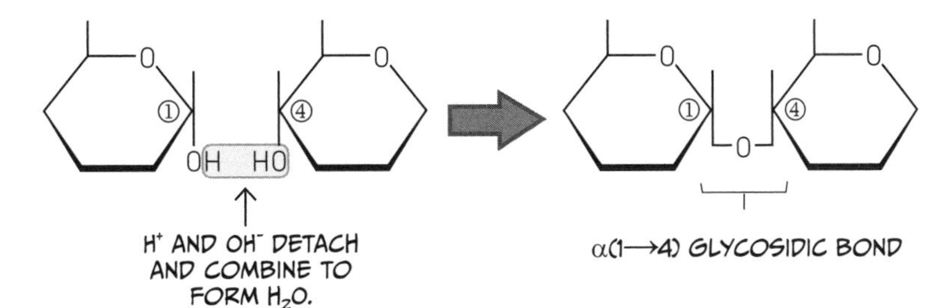

Okay, so the α(1→6) glycosidic bond means that the first carbon of one glucose is attached to the sixth carbon of the next glucose in exactly the same way, right?

Yup, but when the carbon atoms at the first and sixth positions are attached, the glucose molecules can't be side-by-side in a straight line.

* The glucose molecules are simplified throughout this section in order to highlight the relevant bonds.

They can connect only as you see in this diagram—vertically, rather than horizontally.

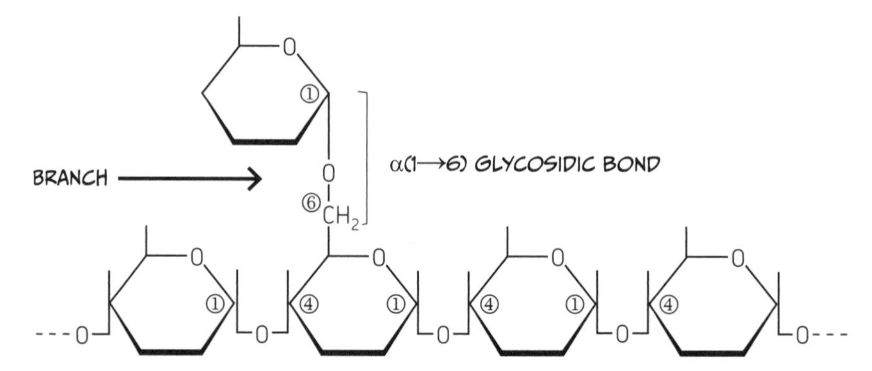

Oh! So that's why a branch happens at that point, and that's what makes it springy. We're really getting to the bottom of things! ♪

By the way, there's another type of bond that connects carbons in glucose. It's called the *β(1→4) glycosidic bond*.

Beta? What's the difference?

Beta means no starch! When glucose molecules are connected via β(1→4) glycosidic bonds, the polysaccharide *cellulose* is formed rather than starch. This is the main component for creating cell walls in plants, and it's also a type of dietary fiber.

Oh yeah! I read about dietary fiber in the latest issue of *Dieter's Digest*. It's difficult to digest, so it just passes right through the body. Heh heh.

That's correct. But dietary fiber also includes substances that easily dissolve in water, such as hemicellulose or pectin, and these are easily digested.

Weird! Are they energy sources for us too?

They are, but let's get back to cellulose, okay? There's an enzyme contained in our saliva called α-amylase, which can break down starches, like rice, into pieces, but α-amylase cannot break down cellulose.

Why not?

Well, look at the β(1→4) glycosidic bond shown in the following figure.

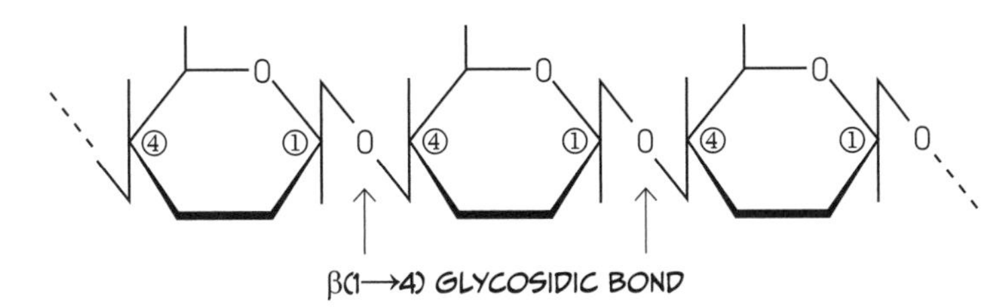

Hey, the parts that are connected are different, aren't they? They have a strange shape, almost like the letter N.

The β(1→4) glycosidic bond differs from the α(1→4) glycosidic bond in that the positions of the hydrogen atom (H) and the hydroxyl group (OH) are flipped around their carbon. This creates a bond that is N-shaped rather than U-shaped, like the α(1→4) glycosidic bond.

Yeah, no offense to the β type, but its connection seems twisted and totally weird!

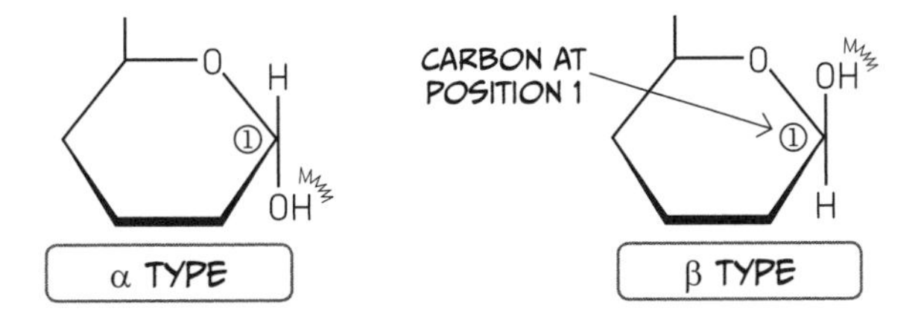

The Greek letters α and β represent the position of the hydroxyl group (OH) on carbon 1. When OH is on the bottom, as shown on the left in the above figure, it is the α *type*. When it's at the top, it is the β *type*.

Because of this difference in structure, α-amylase, which can break down α(1→4) glycosidic bonds, is unable to break down β(1→4) glycosidic bonds.

α-AMYLASE

CAN BREAK THIS DOWN!

CELLULOSE β(1→4)

STARCH α(1→4)

Wow! That twisted connection really makes a big difference!

So cellulose, which is formed by β(1→4) glycosidic bonds, is not broken down in our digestive system. This makes it very effective as dietary fiber. It...you know... *keeps you regular.*

Well. Moving right along! Remember when we were talking about sucrose in fruit? We discovered that sucrose is made up of one glucose and one fructose connected together. The carbon at position 1 of glucose is connected to the carbon at position 2 of fructose, as you can see below.

GLUCOSE

FRUCTOSE

SUCROSE

α(1→2) GLYCOSIDIC BOND

I get it! So sucrose is formed with an α(1→2) glycosidic bond, right?

Exactly! And now that you know how individual monosaccharides are connected together, you should have a much better understanding of why these substances have their unique physical properties.

WHY ARE MOCHI RICE CAKES SPRINGY?

- The secret of why mochi rice cakes are springy is in the structure of the starch in mochi rice.
- The starch of mochi rice contains only amylopectin and does not contain any amylose.
- Since amylopectin uses a connection method called the α(1→6) glycosidic bond, it is a large, branched polysaccharide. It becomes springy because of this branched structure. ♪

WELL, WE SOLVED ALL THE PROFESSOR'S MYSTERIES!
OH YEAH!
生化学
秘
ノート
WOOOOOOOOOOO!
TODAY WAS A LOT OF FUN. PLUS WE GOT A LOT DONE!
(AND WE WERE ABLE TO BE TOGETHER...)
MY TEACHING METHODS MAY NOT BE AS GREAT AS THE PROFESSOR'S, BUT–
モゴ モゴ
!
THAT'S NOT TRUE!
TODAY'S LESSON WAS EASY TO UNDERSTAND AND EVEN PRETTY INTERESTING!
IN FACT, NEMOTO, I...
GUH?

...I THOUGHT YOU WERE INCREDIBLE!
SWOON
ほっ

WELL, IT WAS NOTHING...
REALLY...
MELT
カーッ
キリッ
OKAY! LET'S WRITE UP OUR REPORT RIGHT AWAY!
THUNK
生化学
秘

* KUROSAKA LABS
坂研究室*
I WONDER HOW MY LITTLE LOVEBIRDS ARE DOING.
NEMOTO'S SO LUCKY I'M DOING THIS FOR HIM! WITHOUT MY HELP, THEY'D NEVER END UP...
TOGETHER...
FOREVER!
SQUEE!

ENZYMES ARE THE KEYS TO CHEMICAL REACTIONS

1. Enzymes and Proteins

GROWN A LITTLE?
ピクッ
ARE YOU SAYING I'VE GOTTEN FATTER?!
わあああん
KUMI...
UH, WELL, I GUESS WE SHOULD START THE LESSON.
OOPS
YOU HAVEN'T GROWN AT ALL, I PROMISE.
THE ROLES OF PROTEINS
SO FAR, OF THE THREE NUTRIENTS,
WE'VE LEARNED ABOUT SACCHARIDES AND LIPIDS.
PROTEIN
SACCHARIDE
LIPID
TODAY, I'LL TALK ABOUT THE REMAINING ONE: PROTEINS. YOU CAN'T REALLY STUDY BIOCHEMISTRY WITHOUT KNOWING SOME BASICS ABOUT PROTEINS. THEY'RE THAT IMPORTANT!

DO YOU REMEMBER WHEN WE TALKED ABOUT PROTEIN SYNTHESIS IN OUR VERY FIRST LESSON?
1 PROTEIN SYNTHESIS
2 METABOLISM
3 ENERGY PRODUCTION
4 PHOTOSYNTHESIS
YEAH!
PROTEINS HAVE VERY IMPORTANT ROLES IN KEEPING OUR CELLS ALIVE.
WHAT IMPORTANT ROLES DO PROTEINS PLAY IN OUR BODIES?
SQUEAK
SQUEAK
LET'S LIST THE MAJOR ONES.

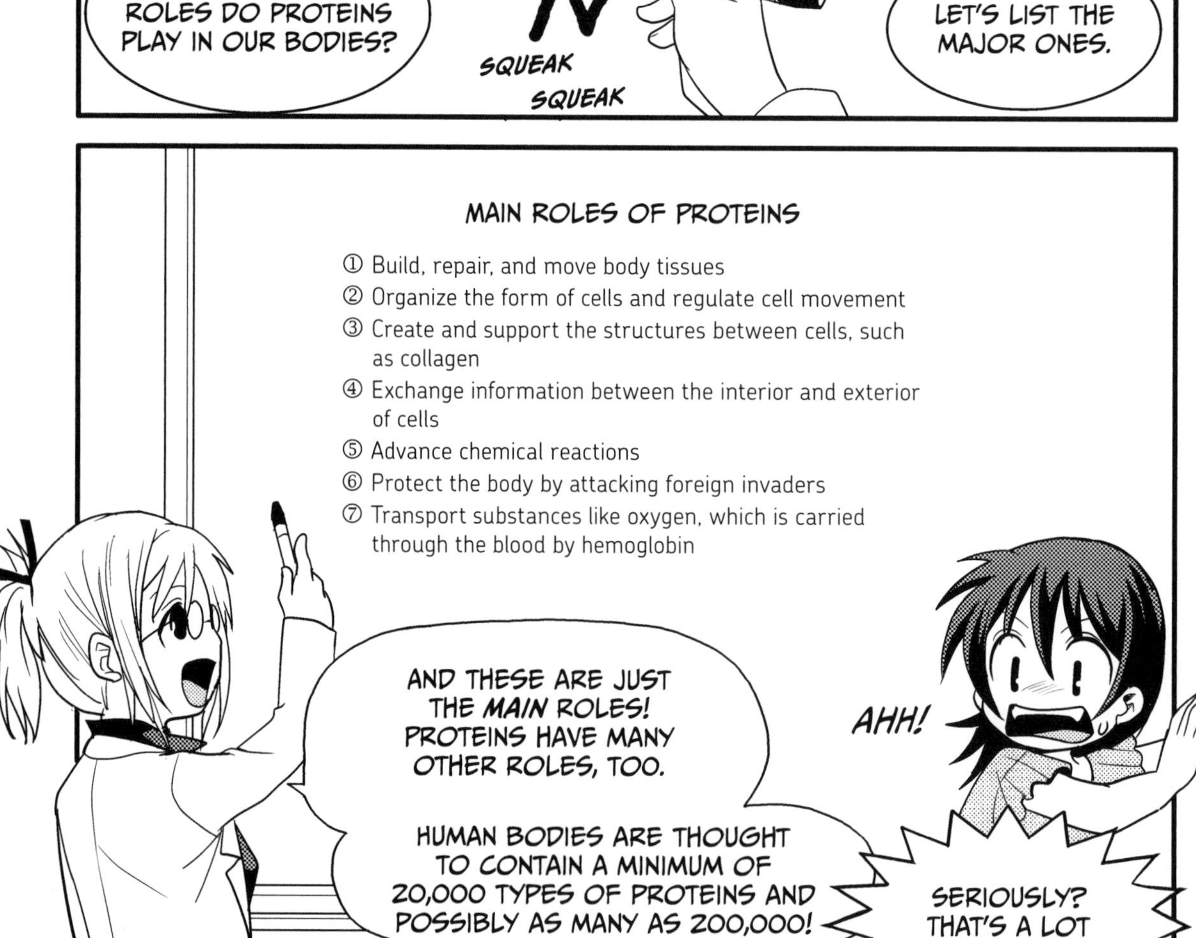
MAIN ROLES OF PROTEINS
① Build, repair, and move body tissues
② Organize the form of cells and regulate cell movement
③ Create and support the structures between cells, such as collagen
④ Exchange information between the interior and exterior of cells
⑤ Advance chemical reactions
⑥ Protect the body by attacking foreign invaders
⑦ Transport substances like oxygen, which is carried through the blood by hemoglobin
AND THESE ARE JUST THE *MAIN* ROLES! PROTEINS HAVE MANY OTHER ROLES, TOO.
HUMAN BODIES ARE THOUGHT TO CONTAIN A MINIMUM OF 20,000 TYPES OF PROTEINS AND POSSIBLY AS MANY AS 200,000!
AHH!
SERIOUSLY? THAT'S A LOT OF PROTEINS!

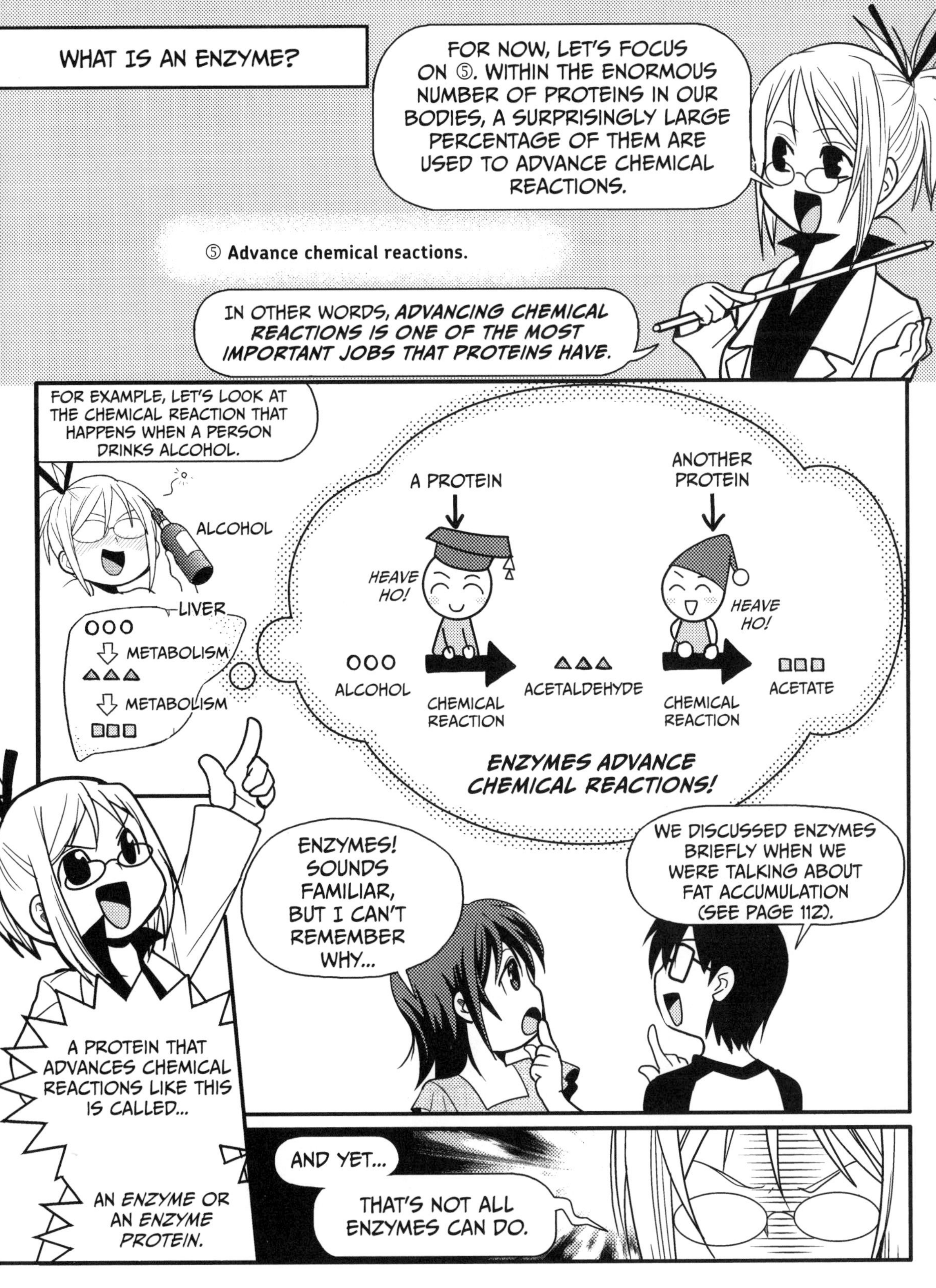
WHAT IS AN ENZYME?
FOR NOW, LET'S FOCUS ON ⑤. WITHIN THE ENORMOUS NUMBER OF PROTEINS IN OUR BODIES, A SURPRISINGLY LARGE PERCENTAGE OF THEM ARE USED TO ADVANCE CHEMICAL REACTIONS.
⑤ Advance chemical reactions.
IN OTHER WORDS, ADVANCING CHEMICAL REACTIONS IS ONE OF THE MOST IMPORTANT JOBS THAT PROTEINS HAVE.
FOR EXAMPLE, LET'S LOOK AT THE CHEMICAL REACTION THAT HAPPENS WHEN A PERSON DRINKS ALCOHOL.
ALCOHOL
LIVER
METABOLISM
METABOLISM
A PROTEIN
ANOTHER PROTEIN
HEAVE HO!
HEAVE HO!
ALCOHOL
CHEMICAL REACTION
ACETALDEHYDE
CHEMICAL REACTION
ACETATE
ENZYMES ADVANCE CHEMICAL REACTIONS!
A PROTEIN THAT ADVANCES CHEMICAL REACTIONS LIKE THIS IS CALLED...
AN ENZYME OR AN ENZYME PROTEIN.
ENZYMES! SOUNDS FAMILIAR, BUT I CAN'T REMEMBER WHY...
WE DISCUSSED ENZYMES BRIEFLY WHEN WE WERE TALKING ABOUT FAT ACCUMULATION (SEE PAGE 112).
AND YET...
THAT'S NOT ALL ENZYMES CAN DO.

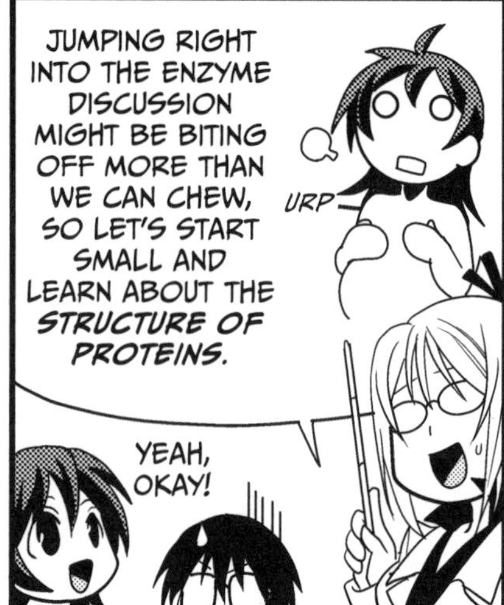

PROTEINS ARE FORMED FROM AMINO ACIDS

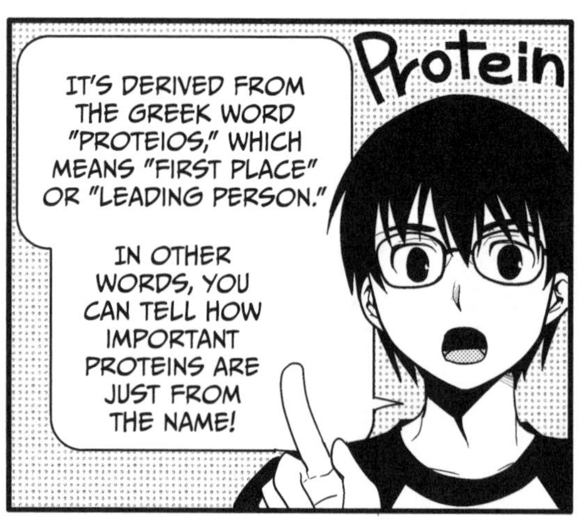

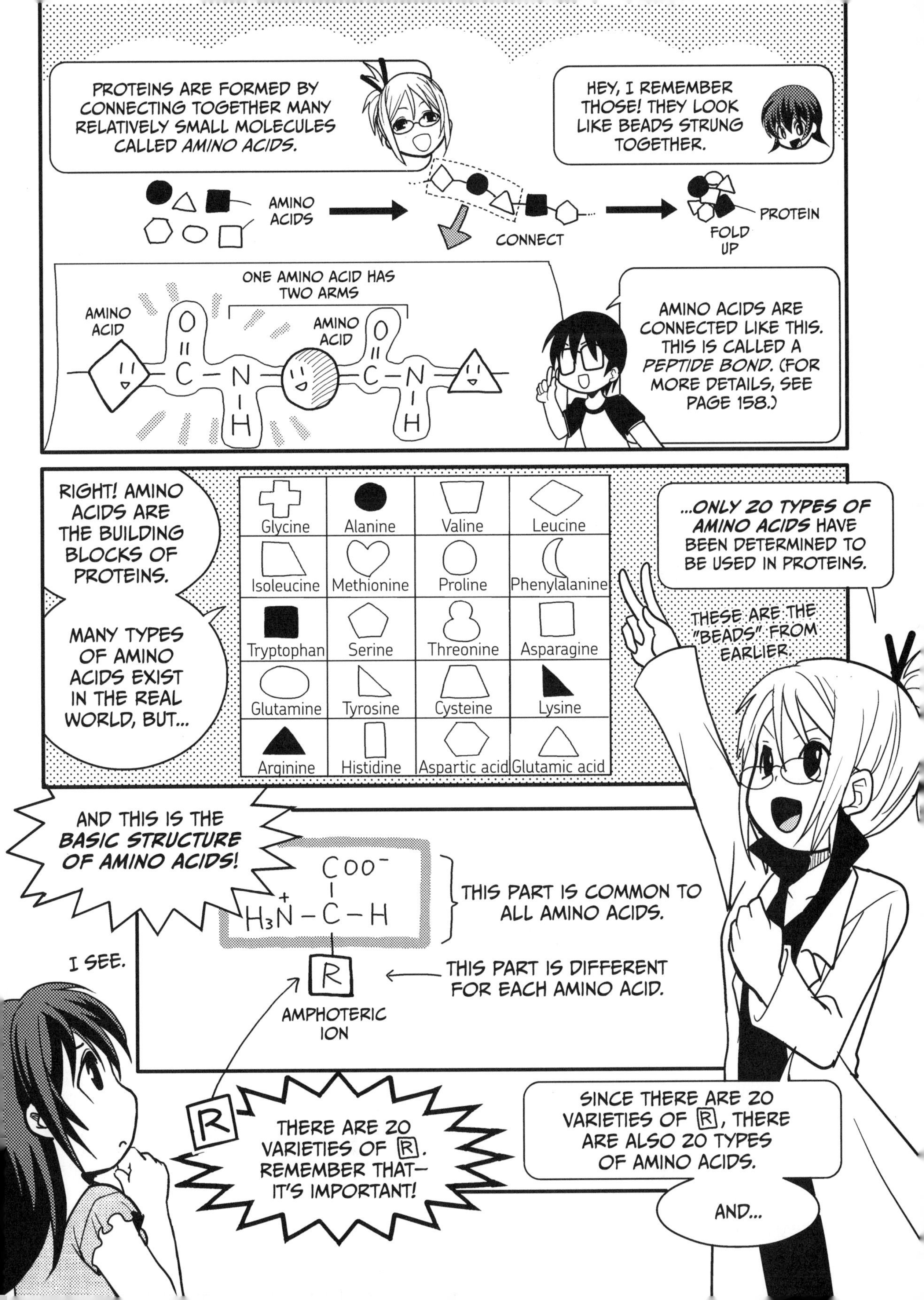

PROTEINS ARE FORMED BY CONNECTING TOGETHER MANY RELATIVELY SMALL MOLECULES CALLED *AMINO ACIDS*.
HEY, I REMEMBER THOSE! THEY LOOK LIKE BEADS STRUNG TOGETHER.
AMINO ACIDS
CONNECT
FOLD UP
PROTEIN
ONE AMINO ACID HAS TWO ARMS
AMINO ACID
AMINO ACID
O
C
N
H
O
C
N
H
AMINO ACIDS ARE CONNECTED LIKE THIS. THIS IS CALLED A *PEPTIDE BOND*. (FOR MORE DETAILS, SEE PAGE 158.)
RIGHT! AMINO ACIDS ARE THE BUILDING BLOCKS OF PROTEINS.
MANY TYPES OF AMINO ACIDS EXIST IN THE REAL WORLD, BUT...
Glycine
Alanine
Valine
Leucine
Isoleucine
Methionine
Proline
Phenylalanine
Tryptophan
Serine
Threonine
Asparagine
Glutamine
Tyrosine
Cysteine
Lysine
Arginine
Histidine
Aspartic acid
Glutamic acid
...*ONLY 20 TYPES OF AMINO ACIDS* HAVE BEEN DETERMINED TO BE USED IN PROTEINS.
THESE ARE THE "BEADS" FROM EARLIER.
AND THIS IS THE *BASIC STRUCTURE OF AMINO ACIDS*!
COO^-
$H_3\overset{+}{N}-C-H$
R
THIS PART IS COMMON TO ALL AMINO ACIDS.
THIS PART IS DIFFERENT FOR EACH AMINO ACID.
AMPHOTERIC ION
I SEE.
R
THERE ARE 20 VARIETIES OF R. REMEMBER THAT—IT'S IMPORTANT!
SINCE THERE ARE 20 VARIETIES OF R, THERE ARE ALSO 20 TYPES OF AMINO ACIDS.
AND...

AMINO ACIDS

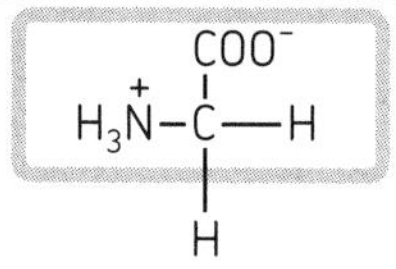

Glycine Gly G	Alanine Ala A	Valine Val V	Leucine Leu L
Isoleucine Ile I	Methionine Met M	Proline Pro P	Phenylalanine Phe F
Tryptophan Trp W	Serine Ser S	Threonine Thr T	Asparagine Asn N
Glutamine Gln Q	Tyrosine Tyr Y	Cysteine Cys C	Lysine Lys K
Arginine Arg R	Histidine His H	Aspartic Acid Asp D	Glutamic Acid Glu E

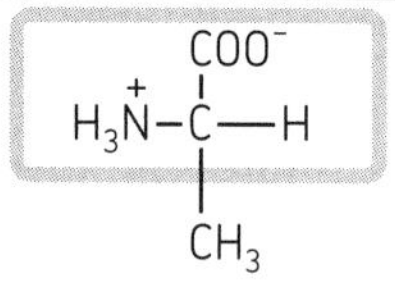

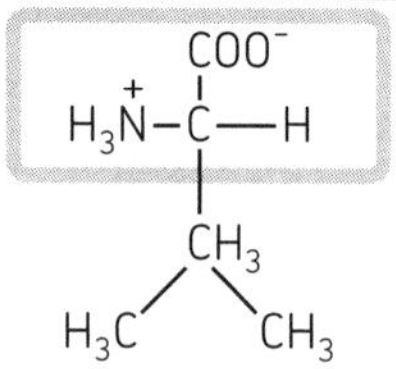

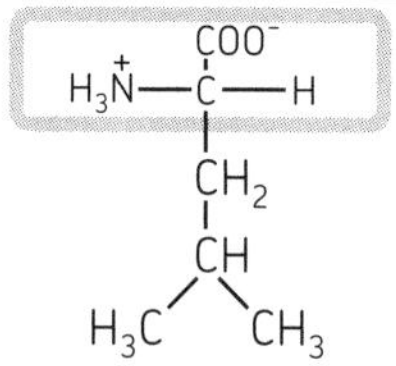

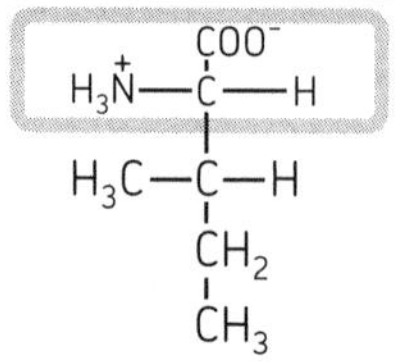

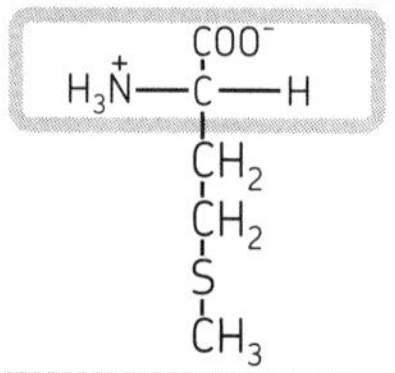

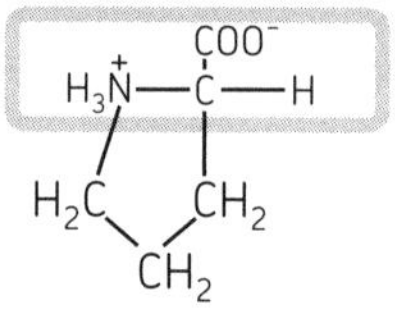

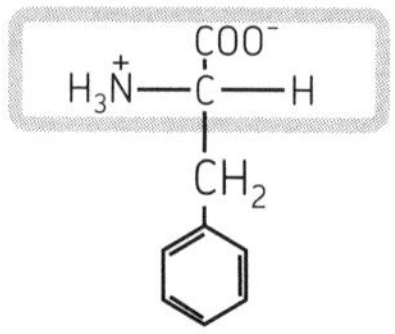

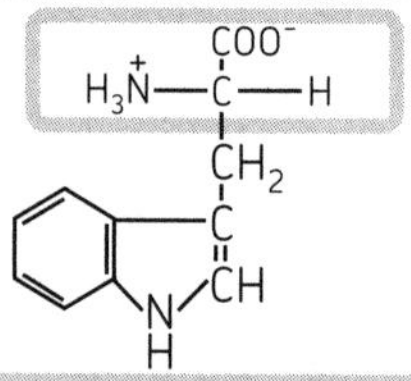

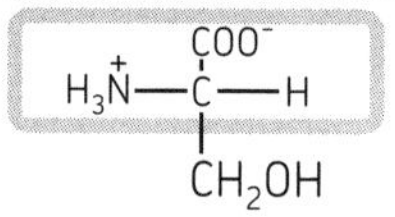

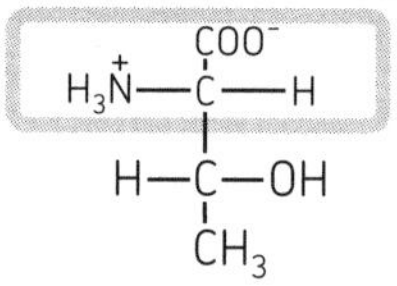

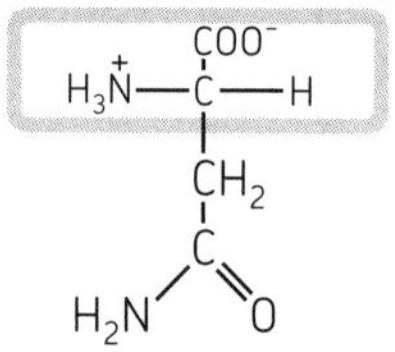

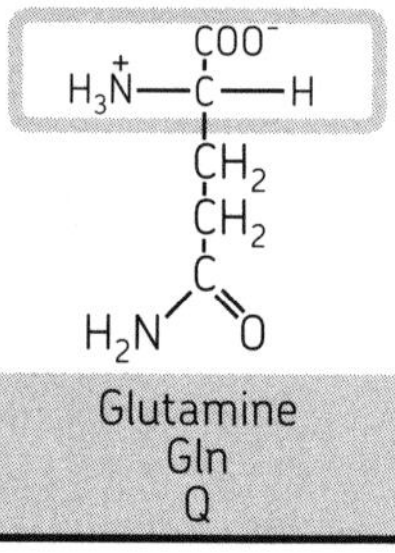

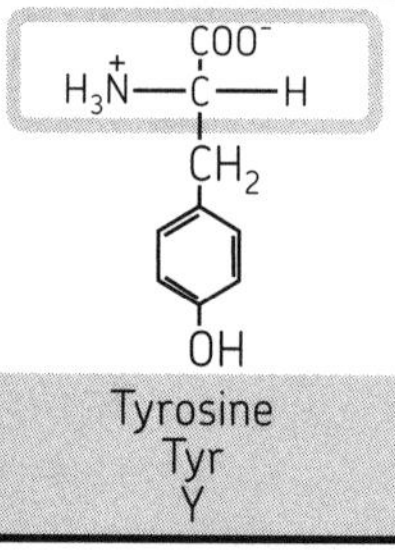

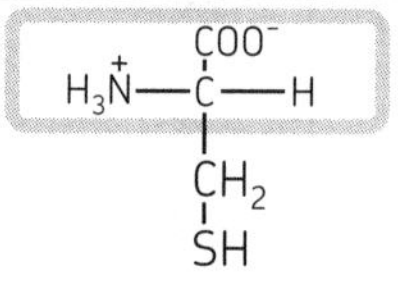

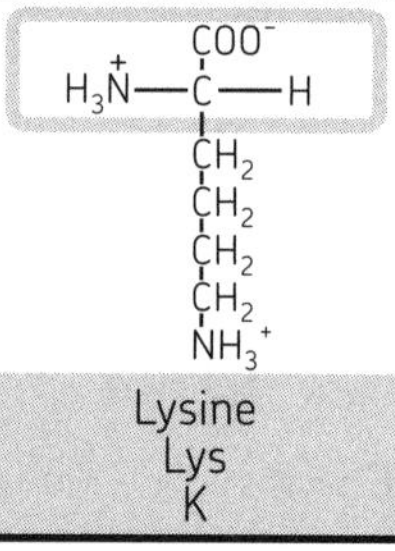

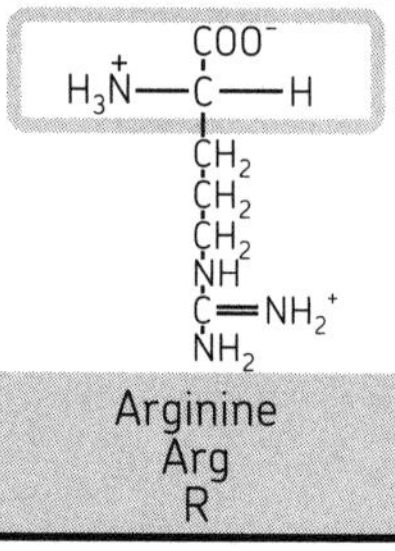

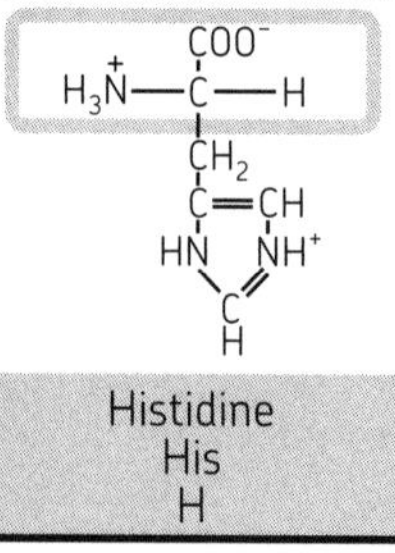

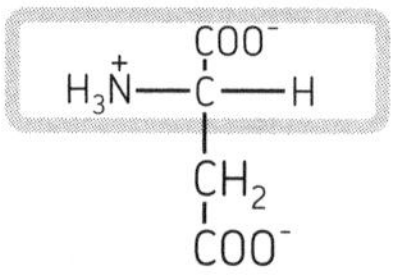

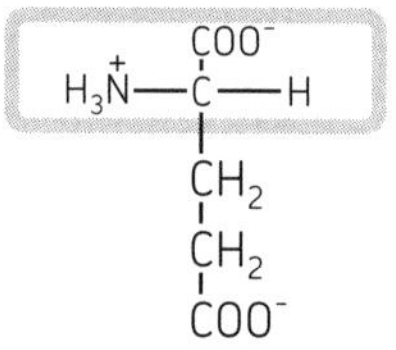

...HERE ARE THE STRUCTURAL FORMULAS OF THOSE 20 AMINO ACIDS!
SWOON
HEY, RELAX! YOU DON'T HAVE TO MEMORIZE THEM ALL RIGHT AWAY!

THE TYPES OF PROTEINS THAT CAN BE CREATED ARE DETERMINED SOLELY BY THESE 20 AMINO ACIDS.
DIFFERENT ORDERS AND DIFFERENT AMOUNTS OF THESE ACIDS PRODUCE DIFFERENT PROTEINS!
ふふふふ
HMM, I GUESS EVEN WITH ONLY 20 KINDS OF BEADS, I CAN MAKE A TON OF DIFFERENT NECKLACES BY COMBINING THEM IN DIFFERENT WAYS.
JANGLE
JANGLE
TRUE, BUT AMINO ACIDS DIFFER FROM BEADS.
FOR ONE, YOU CAN'T JUST STRING TOGETHER A BUNCH OF AMINO ACIDS AND EXPECT TO GET A PROTEIN.
WHAAA?
FOR A PROTEIN TO FUNCTION, IT MUST BE FOLDED INTO AN APPROPRIATE SHAPE,
AND IT HAS TO BE FOLDED IN THE RIGHT ORDER AS WELL.
LET'S TAKE A CLOSER LOOK!

PRIMARY STRUCTURE OF A PROTEIN

First, a long amino acid chain is formed by connecting amino acids one at a time with *peptide bonds*. Remember earlier when we said that amino acids are joined together via chemical reactions to form proteins? This chemical reaction, which creates the peptide bonds, is called the *peptidyl transfer*.

This reaction occurs inside *ribosomes*, which are protein synthesis apparatuses contained in the cytoplasm and attached to the rough endoplasmic reticulum. We'll talk more about ribosomes in Chapter 5.

The long amino acid chain connected together by peptide bonds is called a *polypeptide chain*. This amino acid string is called the *primary structure* of the protein.

AMINO ACID

POLYPEPTIDE CHAIN

H_2N ····· –COOH

ZOOM! IN ON THE PROCESS IN WHICH AMINO ACIDS ARE CONNECTED BY PEPTIDE BONDS!

SYNTHESIZED AMINO ACID CHAIN

H_2O

PEPTIDE BOND

PRIMARY STRUCTURE

SECONDARY STRUCTURE OF A PROTEIN

Each of the 20 amino acids has a characteristic part (indicated by [R] on page 155) that is unique. This is called the *side chain*.

Since a number of difference forces can act on these side chains, such as hydrogen bonds and hydrophobic and electrostatic interactions, neighboring amino acids are attracted to or repelled by each other in particular ways, which results in a characteristic, local three-dimensional structure. This is called the *secondary structure* of the protein.

There are a number of different secondary structures, such as an α-helix, in which a part of the polypeptide chain forms a spiral, and a β-sheet, which takes a folded planar shape.

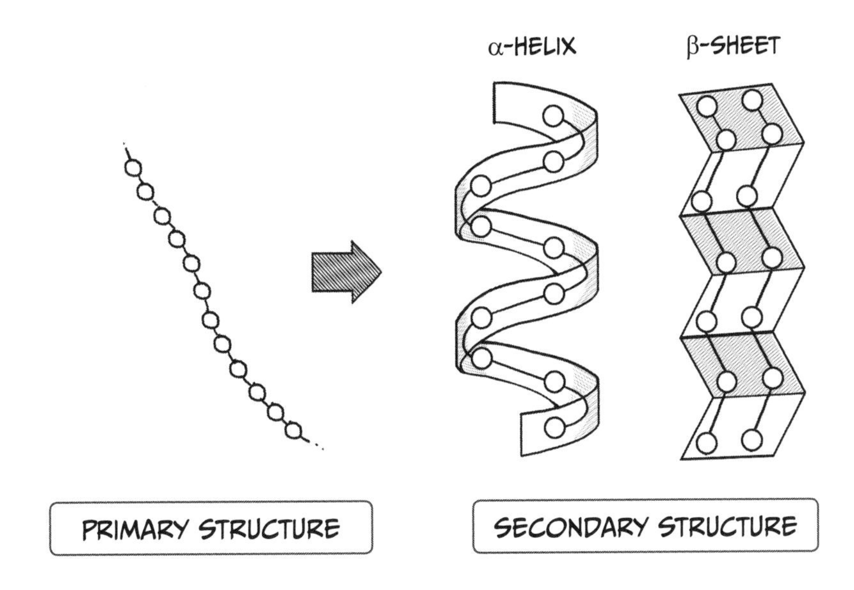

TERTIARY STRUCTURE OF A PROTEIN

Even when the polypeptide chain takes its secondary structure, it's not yet a fully folded, functional protein.

To become functional, the polypeptide chain has to take on a specific three-dimensional shape, which is determined by the interactions of the amino acid side chains. This shape is called the *tertiary structure* of the protein.

For example, myoglobin, which is shown in the following figure, is a type of protein that exists in the muscles of animals. This protein is formed from eight α-helices surrounding an iron-containing *heme*, which binds oxygen.

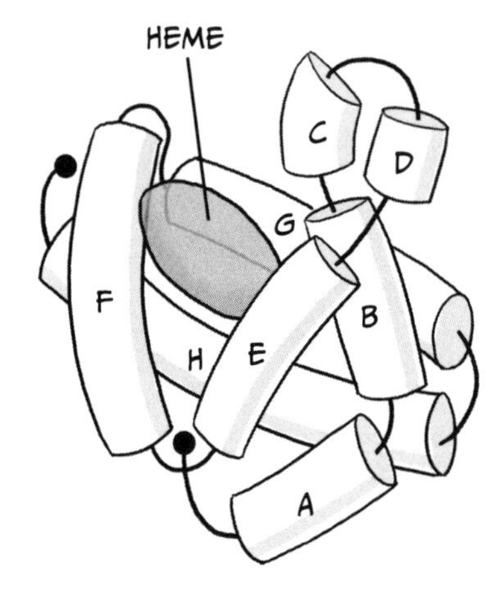

MYOGLOBIN CONSISTS OF THE EIGHT α-HELICES INDICATED BY THE LETTERS A THROUGH H.

TERTIARY STRUCTURE

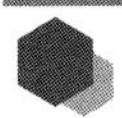

QUATERNARY STRUCTURE OF A PROTEIN AND SUBUNITS

Many proteins are able to operate as a protein or as an enzyme at the tertiary structure stage. However, some proteins create an aggregation in which multiple polypeptide chains that have taken tertiary structures are assembled together into a large functional unit.

For example, our red blood cells contain many iron-binding proteins called *hemoglobin*, which are used to transport oxygen. Hemoglobin is formed by assembling four polypeptide chains called *globin* (two each of two types, α and β, which are indicated below as α_1, α_2, β_1, and β_2). An enzyme such as RNA polymerase II, which creates RNA in our cells, is formed by assembling 12 polypeptide chains.

This state is called the *quaternary structure*, and each of the polypeptide chains used to create the quaternary structure is called a *subunit*.

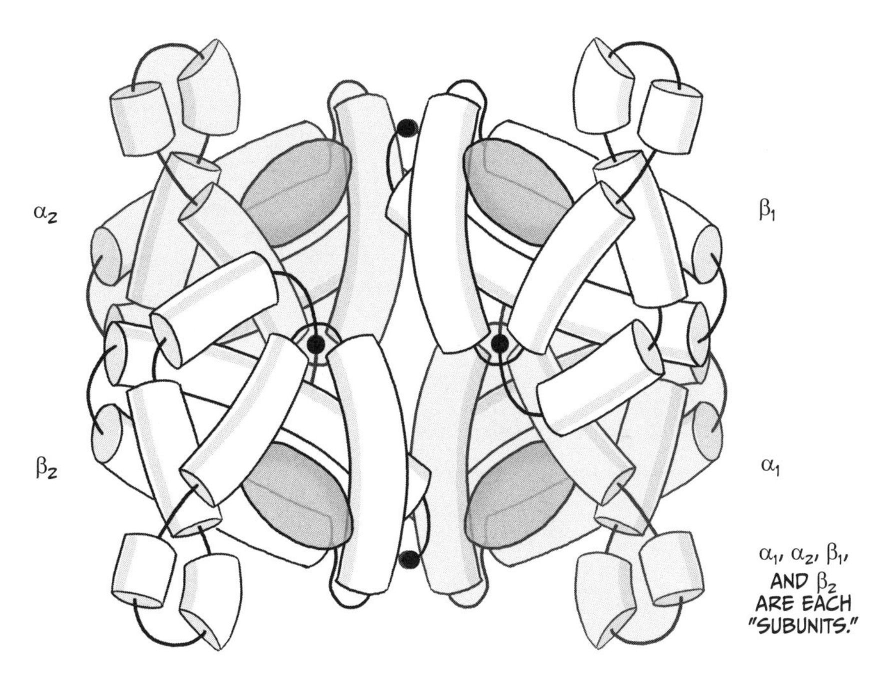

SINCE THE SUBUNITS OF HEMOGLOBIN ARE VERY SIMILAR IN STRUCTURE TO THOSE IN MYOGLOBIN, HEMOGLOBIN IN THIS FIGURE HAS BEEN DRAWN WITH THE SAME STRUCTURE AS ON PAGE 160. HOWEVER, THE STRUCTURES ARE ACTUALLY SOMEWHAT DIFFERENT.

QUATERNARY STRUCTURE

2. An Enzyme's Job

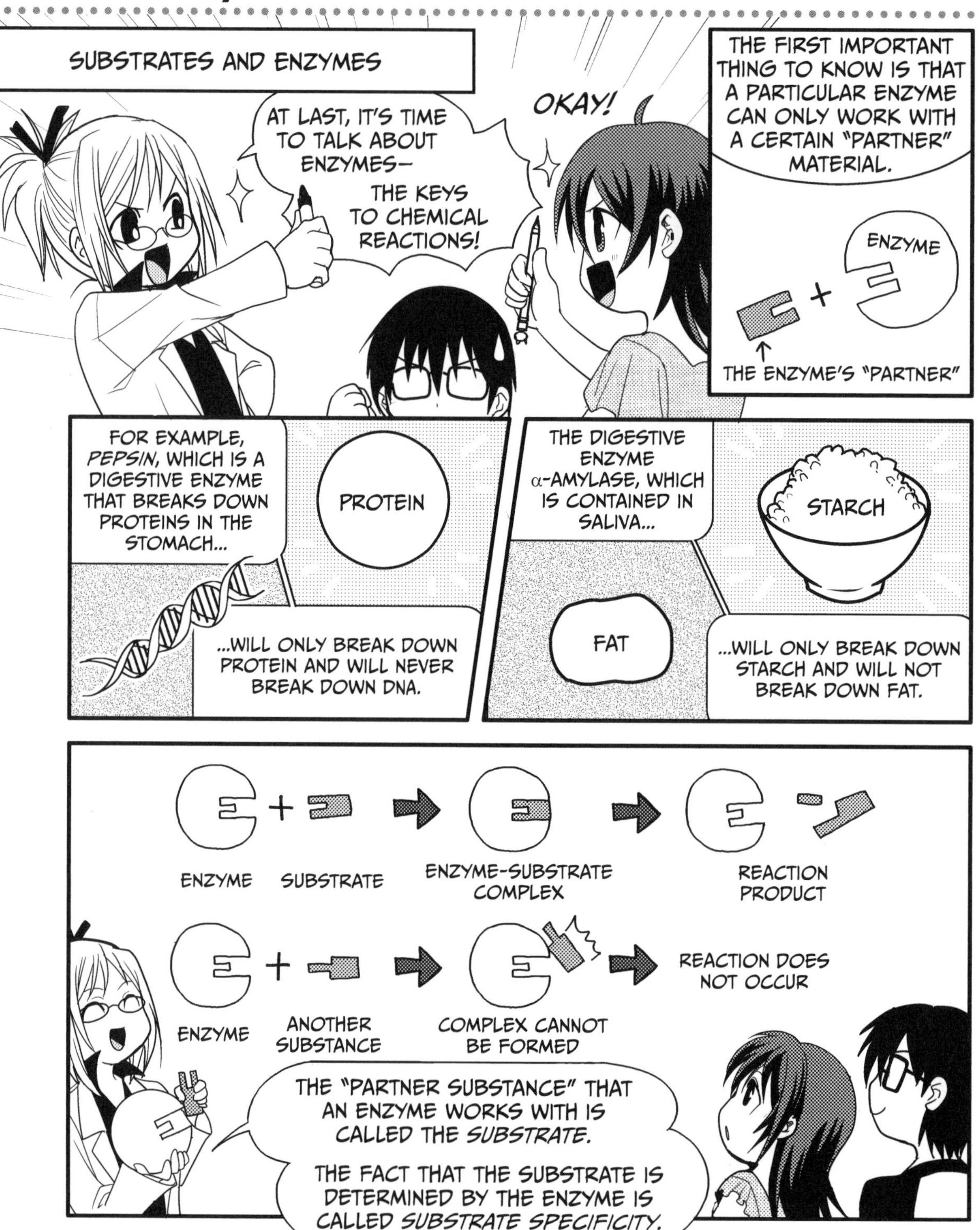

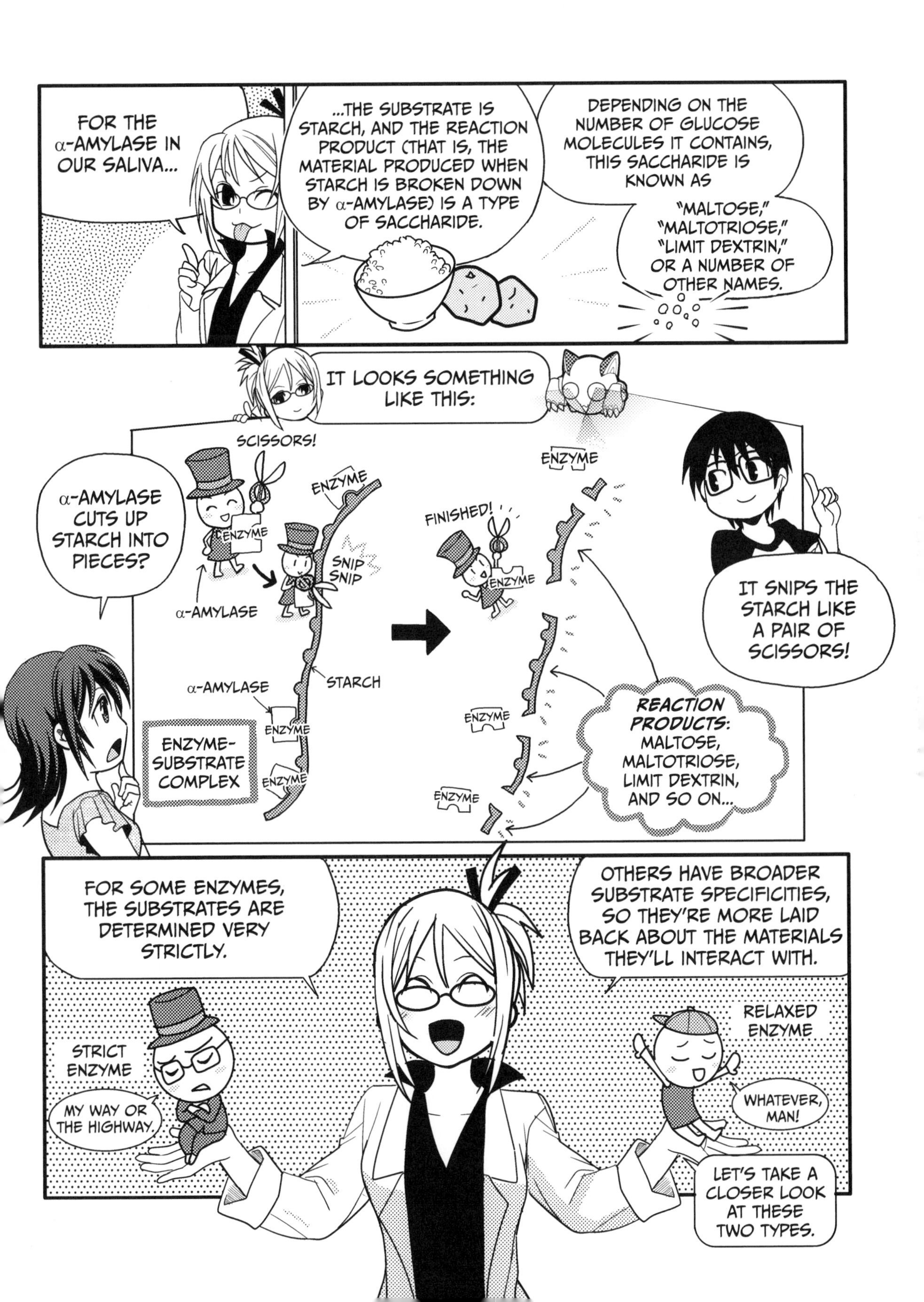
FOR THE α-AMYLASE IN OUR SALIVA...
...THE SUBSTRATE IS STARCH, AND THE REACTION PRODUCT (THAT IS, THE MATERIAL PRODUCED WHEN STARCH IS BROKEN DOWN BY α-AMYLASE) IS A TYPE OF SACCHARIDE.
DEPENDING ON THE NUMBER OF GLUCOSE MOLECULES IT CONTAINS, THIS SACCHARIDE IS KNOWN AS
"MALTOSE," "MALTOTRIOSE," "LIMIT DEXTRIN," OR A NUMBER OF OTHER NAMES.
IT LOOKS SOMETHING LIKE THIS:
α-AMYLASE CUTS UP STARCH INTO PIECES?
SCISSORS!
ENZYME
ENZYME
α-AMYLASE
SNIP SNIP
STARCH
α-AMYLASE
ENZYME
ENZYME
ENZYME-SUBSTRATE COMPLEX
FINISHED!
ENZYME
ENZYME
ENZYME
ENZYME
REACTION PRODUCTS: MALTOSE, MALTOTRIOSE, LIMIT DEXTRIN, AND SO ON...
IT SNIPS THE STARCH LIKE A PAIR OF SCISSORS!
FOR SOME ENZYMES, THE SUBSTRATES ARE DETERMINED VERY STRICTLY.
OTHERS HAVE BROADER SUBSTRATE SPECIFICITIES, SO THEY'RE MORE LAID BACK ABOUT THE MATERIALS THEY'LL INTERACT WITH.
STRICT ENZYME
MY WAY OR THE HIGHWAY.
RELAXED ENZYME
WHATEVER, MAN!
LET'S TAKE A CLOSER LOOK AT THESE TWO TYPES.

CHECK!

STRICT ENZYME? RELAXED ENZYME?

Some enzymes are "strict," which means they only act on very specific substrates. Other enzymes are more "relaxed" and act on a broad range of substrates.

Strict and relaxed? Sounds like the difference between my mom and my dad...

There are enzymes that can act on substances that are similar or closely related to their substrates. Many examples of these enzymes are seen in the digestive system—for example, the *protein catabolism enzymes*.

Remember, there are a huge number of different proteins. If enzymes were *too* specific, there would have to be a separate enzyme to break down every single kind of protein! Things would get pretty unwieldy.

Since proteins are so complex, the enzymes that break them down had to become a bit more flexible.

That's right! Protein catabolism enzymes (the enzymes that break down proteins) often have a certain degree of leeway in the substrates they can interact with.

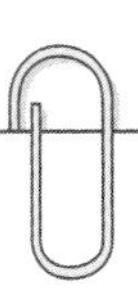

For example, one of the protein catabolism enzymes secreted from the pancreas, carboxypeptidase, detaches amino acids sequentially from the end of a protein.

Carboxypeptidase comes in various types, including A, B, C, and Y. Carboxypeptidase A, for example, can detach almost any amino acid from the C-terminal end of a protein, but it doesn't work well on amino acids with bulky or aromatic R-groups, like arginine, lysine, and proline.

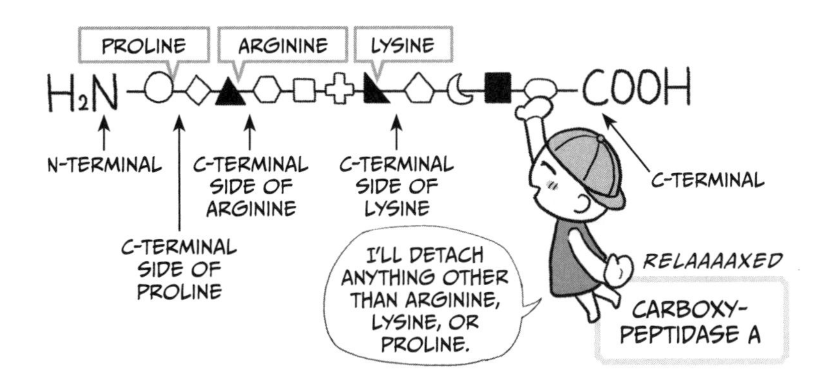

There are 20 types of amino acids that make up proteins, right? So even though carboxypeptidase A can't deal with arginine, lysine, and proline, there are still 17 types it *can* deal with. It seems pretty flexible!

That's right. It has some serious leeway in the substrates it can interact with.

However, there are also some "strict" protein catabolism enzymes. For example, trypsin cuts only through the C-terminal side of arginine and lysine.

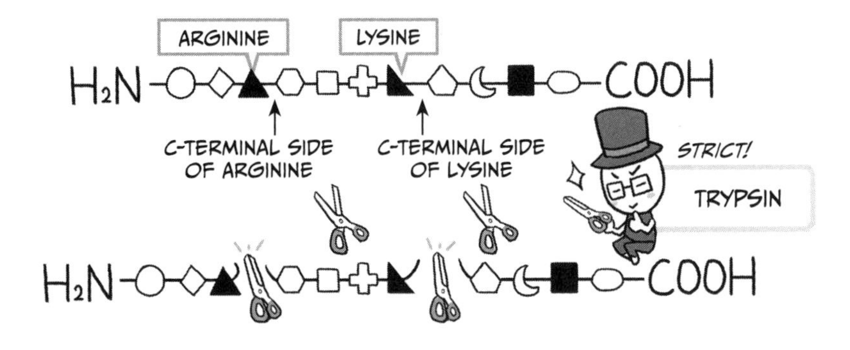

* A NEWLY DISCOVERED ENZYME IS ASSIGNED AN EC NUMBER AND SYSTEM NAME AS WELL AS A RECOMMENDED NAME ACCORDING TO STANDARDS DETERMINED BY THE ENZYME COMMISSION OF THE INTERNATIONAL UNION OF BIOCHEMISTRY AND MOLECULAR BIOLOGY (IUBMB).

NOW FOR THE SIX GROUPS!
Catalytic Reaction Types of Enzymes
EC 1. Oxidoreductases
EC 2. Transferases
EC 3. Hydrolases
EC 4. Lyases
EC 5. Isomerases
EC 6. Ligases
TODAY, WE'LL INTRODUCE EC 2 (TRANSFERASES) AND EC 3 (HYDROLASES) AS REPRESENTATIVES OF THESE CATALYTIC REACTION TYPES.
TODAY'S GUESTS ARE...
WHA?
ALRIGHT, LET'S HAVE THEM MAKE THEIR ENTRANCE!
TRANSFERASE AND HYDROLASE!
TA-DAAAAAH!
TRANSFERASE
HYDROLASE
FUNCTIONAL GROUP
H
OH
EC2.
TRANSFERASE
EC3.
HYDROLASE
NOW LET'S FIND OUT WHAT KIND OF ENZYMES THESE GROUPS CONTAIN AND WHAT JOBS THEY DO.

TRANSFERASES
AN ENZYME FOR TRANSFERRING A PART OF ONE MOLECULE TO ANOTHER IS CALLED A *TRANSFERASE*.
TRANSFERASE
TRANSFERASE IS THE GENERIC NAME FOR ENZYMES THAT TRANSFER A SPECIFIC CLUMP OF ATOMS (FUNCTIONAL GROUP) TO A CHEMICAL COMPOUND OTHER THAN WATER.
TRANSFERASE
SPECIAL DELIVERY!
FUNCTIONAL GROUP
REPRESENTED BY **EC 2.X.X.X**
FOR EXAMPLE, THIS THYMIDYLATE SYNTHASE (EC 2.1.1.45) IS A TRANSFERASE:
5,10-METHYLENE TETRAHYDROFOLATE
H_2N
N
N
N
N
O
H
C
H_2
N
R
TRANSFERASE
URACIL
P
O
H
METHYL GROUP $-CH_3$
THYMINE
CH_3
P
O
IMPORTANT!
THE METHYL GROUP ($-CH_3$) THAT WAS TAKEN FROM A SUBSTANCE CALLED *5,10-METHYLENE TETRAHYDROFOLATE* IS ATTACHED TO URACIL, CHANGING IT INTO THYMINE.
I SEE...
URACIL AND THYMINE ARE INGREDIENTS OF NUCLEIC ACID, BUT WE'LL GET TO THOSE LATER (SEE CHAPTER 5).

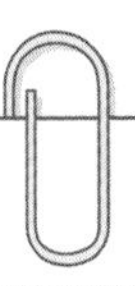

 GLUCOSYLTRANSFERASE DETERMINES BLOOD TYPE

Remember when we solved the blood type mystery (in Chapter 3)?

When we were trying to figure out what determines blood type, we discovered that it was the work of an enzyme.

I remember!

The ABO blood group system is classified according to the differences between the "sugar chains" on the surface of red blood cells. A sugar chain is a collection of monosaccharides connected together!

And the differences between those chains are the monosaccharides at the tips, right?

Yeah! For people with Type A blood, that monosaccharide is *N*-acetylgalactosamine (GalNAc); for people with Type B blood, it's galactose (Gal); and for people with Type O blood, there is no monosaccharide at the tip.

(After writing such a detailed report for the professor, I'll probably remember that for the rest of my life...)

That's correct. The particular monosaccharide that's attached to the tip of a sugar chain (or the lack of one) is determined by a certain *gene*.

Genes are like the "blueprints" for proteins. (Remember: An enzyme is a type of protein.)

So what really determines blood type is the gene that creates *glycosyltransferase*, which is the enzyme that attaches a particular monosaccharide to the tip of a sugar chain found on the surface of the red blood cell.

So glycosyltransferase attaches a certain monosaccharide, and the type of that monosaccharide determines blood type?

That's right. Look again at the structure of the sugar chain of each blood type.

For the sugar chain of people with Type A blood, the tip is GalNAc.

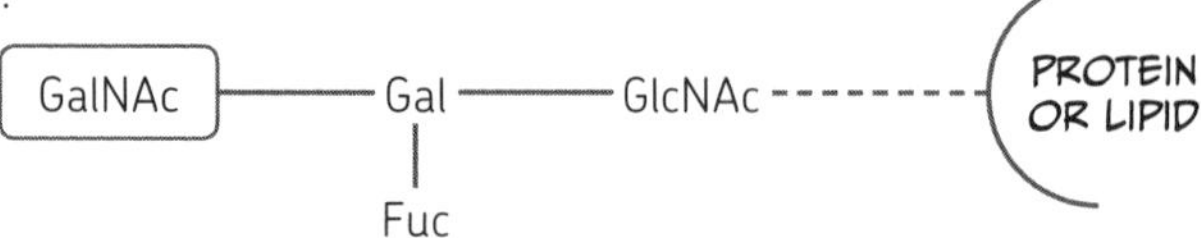

For the sugar chain of people with Type B blood, the tip is Gal.

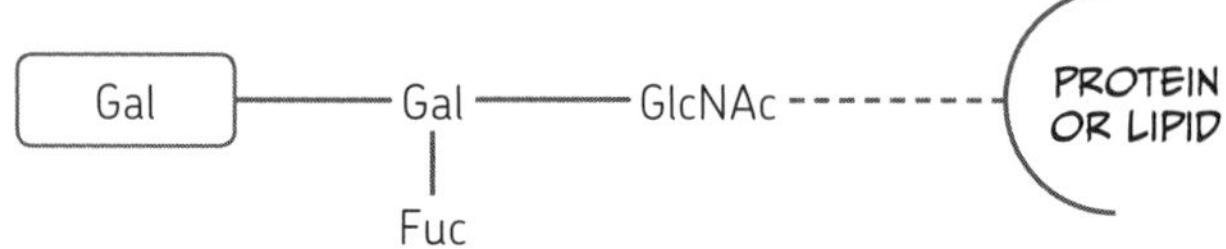

For the sugar chain of people with Type O blood, there is no tip.

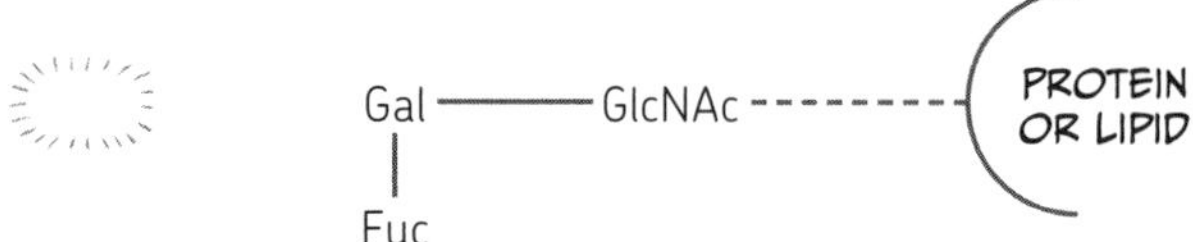

MONOSACCHARIDE NAMES

GalNAc : *N*-acetylgalactosamine
Gal : Galactose
Fuc : Fucose
GlcNAc : *N*-acetylglucosamine

Let's review it again. The only difference in these three types is the monosaccharide at the very tip.

For people with **Type A** blood, it's *N*-acetylgalactosamine.
For people with **Type B** blood, it's galactose.
For people with **Type O** blood, it's nothing!

Yeah, yeah, I've got it.

So, the sugar chain that Type O people have is the "prototype," or the minimal sugar chain. People with A and B types have the O type chain plus something extra.

If someone has the transferase gene that attaches *N*-acetylgalactosamine, his or her blood type will be Type A! But if someone has the transferase gene that attaches galactose, his or her blood type will be Type B!

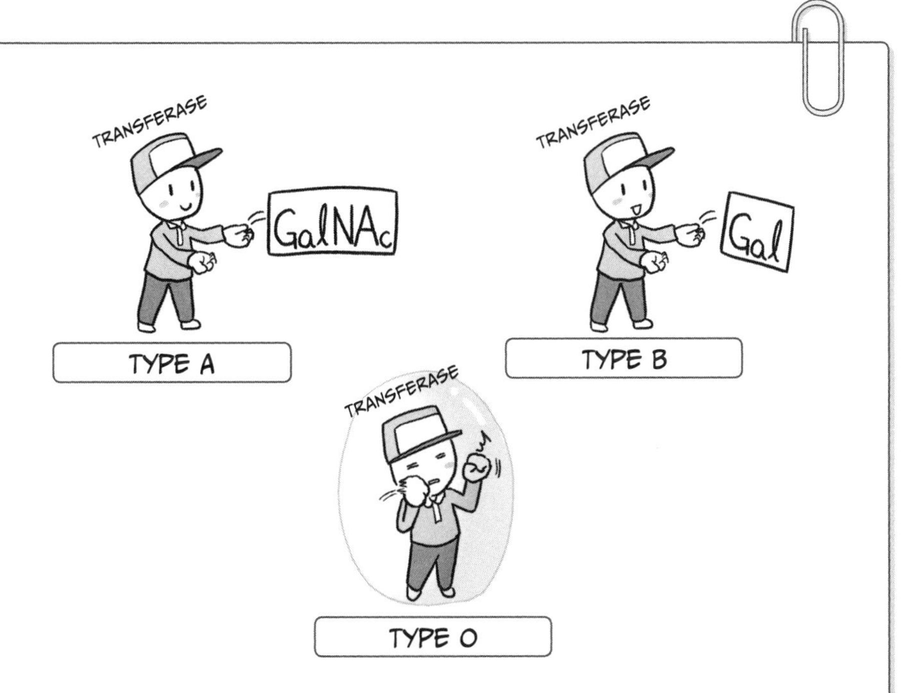

That all makes sense, but I still don't quite understand what's going on with people with Type O blood. Why don't they get monosaccharides on the end of their sugar chains? Seems unfair!

Type O people also have the genes associated with glycosyltransferases, but since there is no enzyme activity for the proteins created by those genes, no saccharide is attached at the tips. The activity may have been lost because of a genetic mutation during the evolutionary process.

A mutation? Cool!

So to wrap it all up: The ABO blood group system is nothing but the result of differences in glycosyltransferase genes.

Nemoto, you must have a genetic mutation that gives you an oversized brain. How else could you possibly know so much?!

Gyuh...

HYDROLASES

HYDROLASE
AS THE NAME SUGGESTS, A *HYDROLASE* IS AN ENZYME THAT USES WATER TO BREAK DOWN ITS SUBSTRATE.
HYDROLASES ARE REPRESENTED BY **EC 3.X.X.X.**
REMEMBER, A WATER MOLECULE IS MADE UP OF TWO HYDROGENS AND ONE OXYGEN: H_2O.
H
OH
A HYDROLASE SPLITS THIS INTO *H* AND *OH*.
IT THEN INSERTS THESE INTO A SUBSTRATE TO BREAK IT DOWN INTO TWO PARTS!
HYDROLASE
BREAK IT UP, GUYS!
H
OH
IT'S LIKE BREAKING UP A FIGHT!
OR SEPARATING TWO PEOPLE IN LOVE...
IF YOU THINK OF THE PARTS OF A SUBSTRATE AS TWO PEOPLE HOLDING HANDS, HYDROLASE SEPARATES THEM BY MAKING ONE HOLD H AND THE OTHER HOLD OH.
HYDROLASE IS WHAT BREAKS DOWN THE COMMON CURRENCY OF ENERGY (ATP)!
α-AMYLASE, WHICH BREAKS DOWN STARCH, AND PEPSIN, WHICH BREAKS DOWN PROTEIN...
PROTEIN
STARCH
...ARE BOTH HYDROLASES!
NOW, LET'S EXAMINE THE WORK OF α-AMYLASE (EC 3.2.1.1)!
THE α-AMYLASE IN OUR SALIVA BREAKS DOWN STARCHES, SUCH AS RICE.

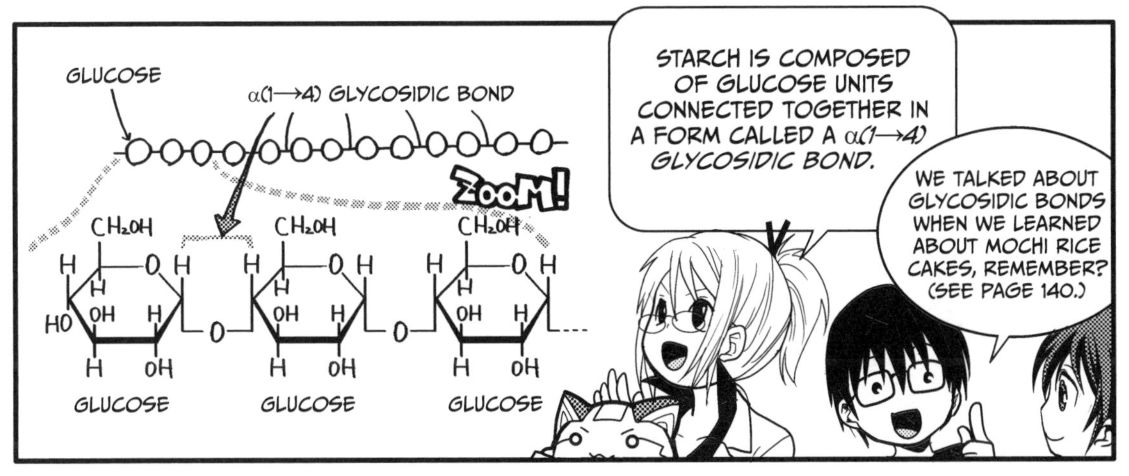
GLUCOSE
α(1→4) GLYCOSIDIC BOND
ZOOM!
CH₂OH
GLUCOSE
GLUCOSE
GLUCOSE
STARCH IS COMPOSED OF GLUCOSE UNITS CONNECTED TOGETHER IN A FORM CALLED A α(1→4) GLYCOSIDIC BOND.
WE TALKED ABOUT GLYCOSIDIC BONDS WHEN WE LEARNED ABOUT MOCHI RICE CAKES, REMEMBER? (SEE PAGE 140.)

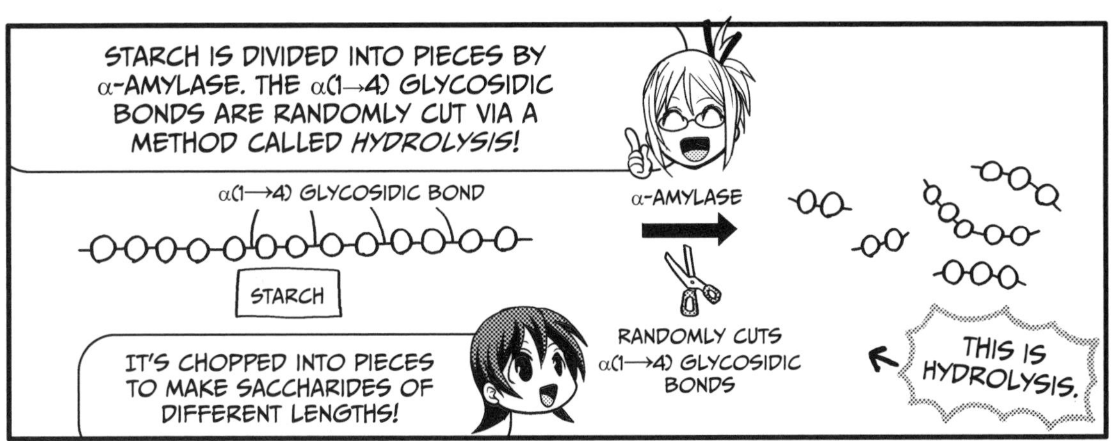
STARCH IS DIVIDED INTO PIECES BY α-AMYLASE. THE α(1→4) GLYCOSIDIC BONDS ARE RANDOMLY CUT VIA A METHOD CALLED *HYDROLYSIS*!
α(1→4) GLYCOSIDIC BOND
STARCH
α-AMYLASE
RANDOMLY CUTS α(1→4) GLYCOSIDIC BONDS
THIS IS HYDROLYSIS.
IT'S CHOPPED INTO PIECES TO MAKE SACCHARIDES OF DIFFERENT LENGTHS!

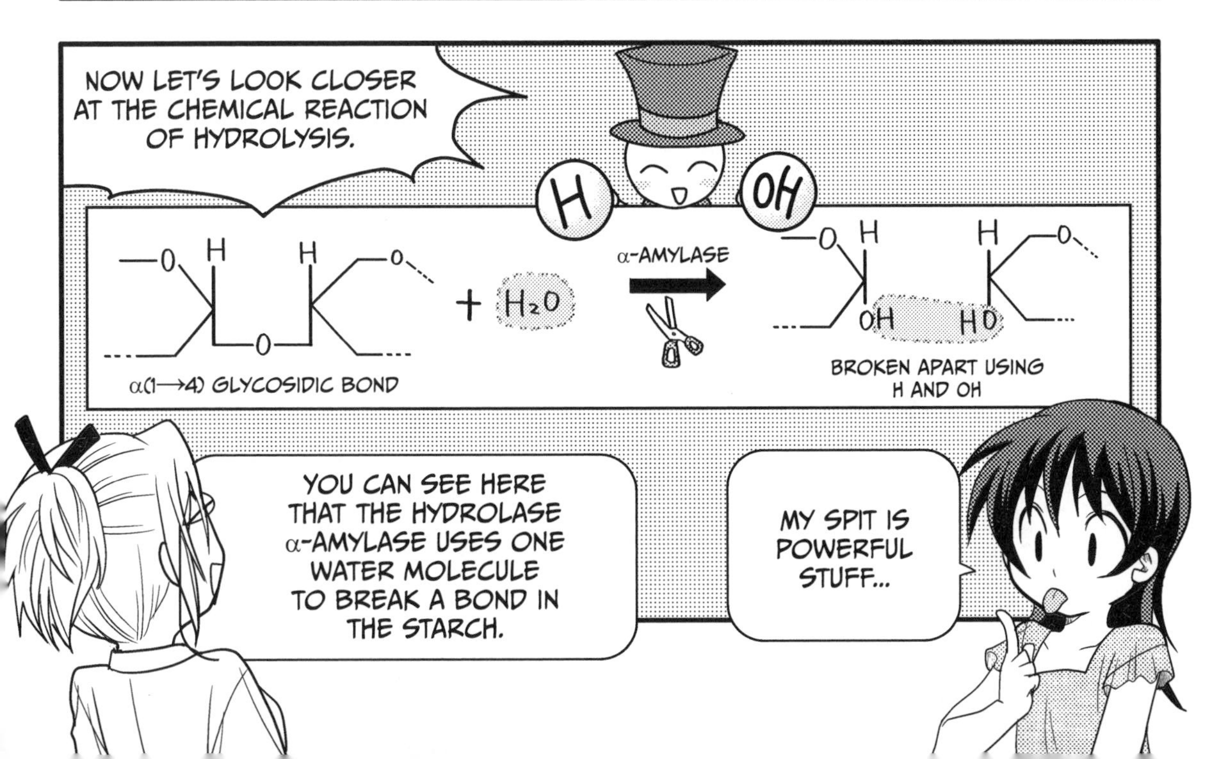
NOW LET'S LOOK CLOSER AT THE CHEMICAL REACTION OF HYDROLYSIS.
H
OH
+ H₂O
α-AMYLASE
α(1→4) GLYCOSIDIC BOND
BROKEN APART USING H AND OH
YOU CAN SEE HERE THAT THE HYDROLASE α-AMYLASE USES ONE WATER MOLECULE TO BREAK A BOND IN THE STARCH.
MY SPIT IS POWERFUL STUFF...

3. Using Graphs to Understand Enzymes

WHY ARE ENZYMES IMPORTANT FOR CHEMICAL REACTIONS?

A substance that expedites a chemical reaction is called a *catalyst*. An enzyme, which is a type of catalyst, is also known as a *biological catalyst*.

Although an enzyme is required to make a chemical reaction advance efficiently, it is not neccesarily required for the chemical reaction to occur.

Most chemical reactions in the body will occur eventually even without enzymes, but many would take an overwhelmingly long time. Complex chemical reactions, like glycolysis, may never reach completion without a catalyst.

Enzymes are important because, for an organism that must maintain specific conditions and produce energy to stay alive, the chemical reactions that occur inside its body must be as efficient as possible. For living organisms as a whole, it would be disastrous if reactions took too long.

THERE WOULD BE NO LIFE AS WE KNOW IT!

BUT BECAUSE ENZYMES EXIST...

EVERYONE'S ALIVE AND HAPPY!

Now we'll use some simple graphs and formulas to study the essential qualities of chemical reactions that rely on enzymes, and we'll learn why enzymes are so meaningful to those reactions.

WHAT IS ACTIVATION ENERGY?

A FIXED AMOUNT OF ENERGY IS REQUIRED FOR A CHEMICAL REACTION TO PROCEED SMOOTHLY. THIS IS CALLED *ACTIVATION ENERGY*.

TAKE A LOOK AT THIS GRAPH, WHICH SHOWS THE PROGRESS OF AN INDIVIDUAL CHEMICAL REACTION.

A : SUBSTRATE

B : REACTION PRODUCT

AMOUNT OF ENERGY

ACTIVATION ENERGY

A

B

REACTION PROGRESS

HERE, CHEMICAL SUBSTRATE, ***SUBSTRATE A***, IS TRANSFORMED VIA A CHEMICAL REACTION TO PRODUCE REACTION ***PRODUCT B***.

FOR THE REACTION TO PRODUCE B FROM A, THE ACTIVATION ENERGY HAS TO BE ADDED TO THE MIX.

SINCE A AND B ARE DIFFERENT SUBSTANCES, THEY HAVE DIFFERENT AMOUNTS OF ENERGY.

AMOUNT OF ENERGY

A

B

LOOK AT THE ENERGY VALUES OF A AND B!

THE ACTIVATION ENERGY DOES NOT AFFECT THE DIFFERENCE BETWEEN THE ENERGY VALUES OF A AND B.

IT'S ALMOST LIKE THE SUBSTRATE HAS TO CLIMB OVER A REALLY HIGH WALL TO ESCAPE IMPRISONMENT AND TRANSFORM INTO THE REACTION PRODUCT.

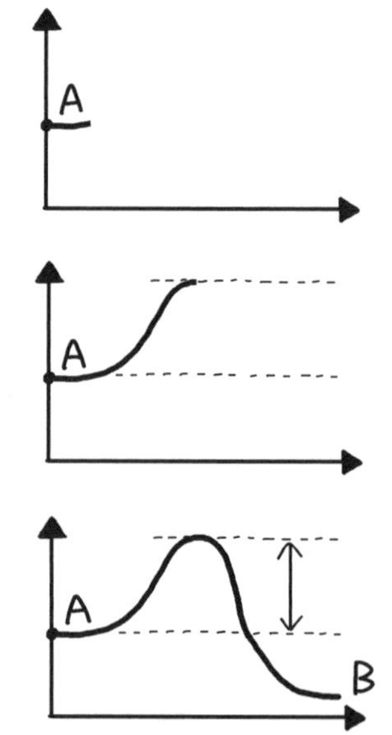

THE HIGHEST PART OF THIS WALL IS CALLED THE *ACTIVATION BARRIER* OR *ENERGY BARRIER*.

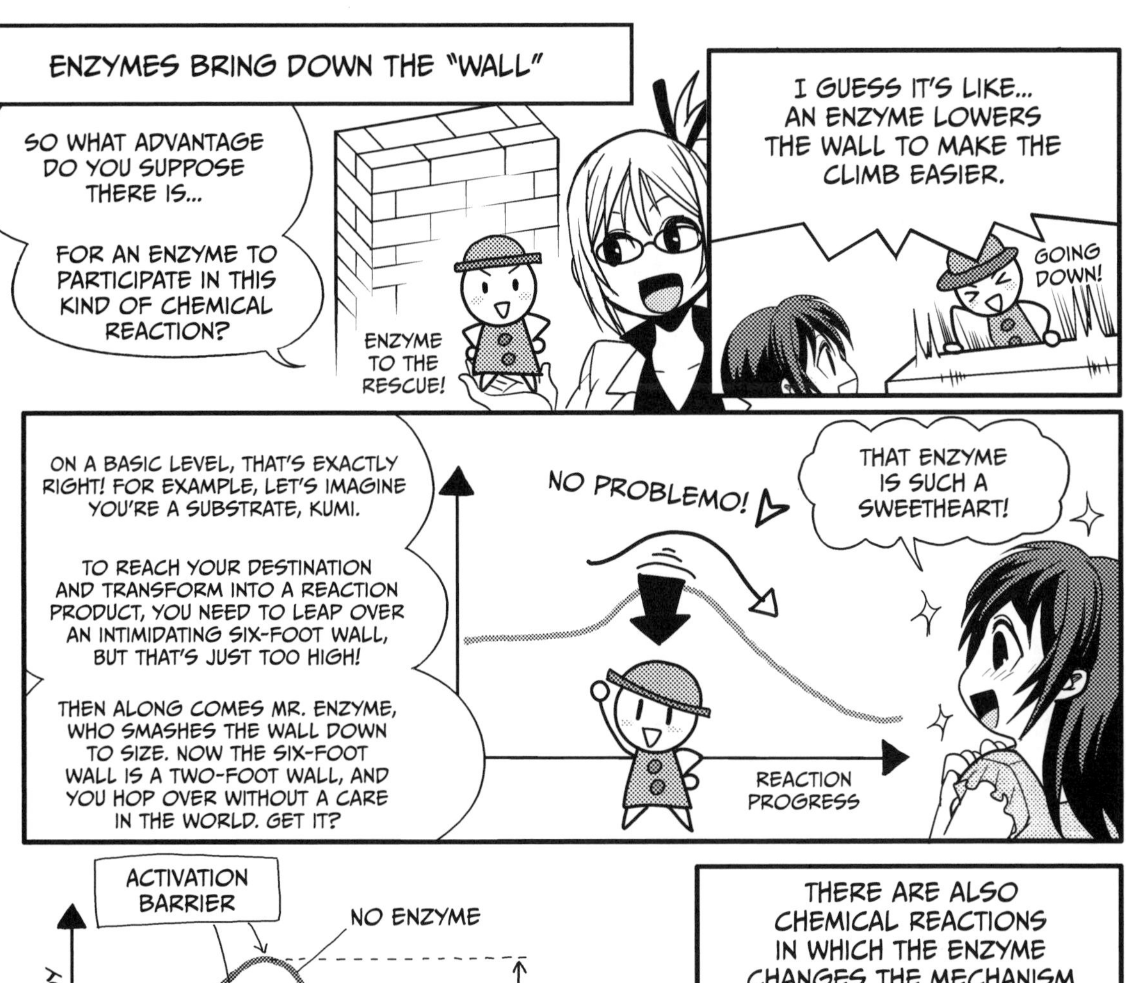
ENZYMES BRING DOWN THE "WALL"
SO WHAT ADVANTAGE DO YOU SUPPOSE THERE IS...
FOR AN ENZYME TO PARTICIPATE IN THIS KIND OF CHEMICAL REACTION?
ENZYME TO THE RESCUE!
I GUESS IT'S LIKE... AN ENZYME LOWERS THE WALL TO MAKE THE CLIMB EASIER.
GOING DOWN!
ON A BASIC LEVEL, THAT'S EXACTLY RIGHT! FOR EXAMPLE, LET'S IMAGINE YOU'RE A SUBSTRATE, KUMI.
TO REACH YOUR DESTINATION AND TRANSFORM INTO A REACTION PRODUCT, YOU NEED TO LEAP OVER AN INTIMIDATING SIX-FOOT WALL, BUT THAT'S JUST TOO HIGH!
THEN ALONG COMES MR. ENZYME, WHO SMASHES THE WALL DOWN TO SIZE. NOW THE SIX-FOOT WALL IS A TWO-FOOT WALL, AND YOU HOP OVER WITHOUT A CARE IN THE WORLD. GET IT?
NO PROBLEMO!
THAT ENZYME IS SUCH A SWEETHEART!
REACTION PROGRESS

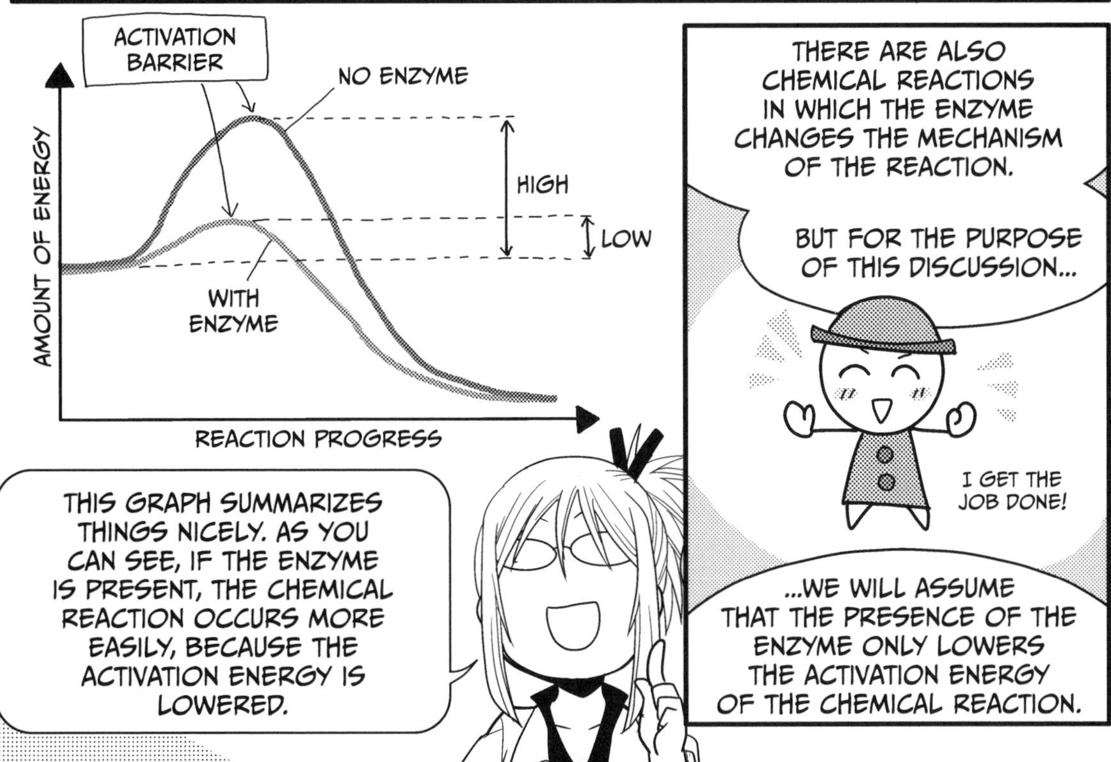
ACTIVATION BARRIER
NO ENZYME
HIGH
LOW
WITH ENZYME
AMOUNT OF ENERGY
REACTION PROGRESS
THIS GRAPH SUMMARIZES THINGS NICELY. AS YOU CAN SEE, IF THE ENZYME IS PRESENT, THE CHEMICAL REACTION OCCURS MORE EASILY, BECAUSE THE ACTIVATION ENERGY IS LOWERED.
THERE ARE ALSO CHEMICAL REACTIONS IN WHICH THE ENZYME CHANGES THE MECHANISM OF THE REACTION.
BUT FOR THE PURPOSE OF THIS DISCUSSION...
I GET THE JOB DONE!
...WE WILL ASSUME THAT THE PRESENCE OF THE ENZYME ONLY LOWERS THE ACTIVATION ENERGY OF THE CHEMICAL REACTION.

MAXIMUM REACTION RATE

THE ABILITY OF AN ENZYME TO ACT ON THE SUBSTRATE TO CREATE THE REACTION PRODUCT IS CALLED *ENZYME ACTIVITY*.

ENZYME POWER!

SUBSTRATE → REACTION PRODUCT

ENZYME ACTIVITY

THE SPEED OF THIS REACTION IS CALLED THE *REACTION RATE*.

* THE *V* STANDS FOR "VELOCITY," WHICH IS JUST ANOTHER WORD FOR "RATE."

V_{max}

WILL NOT INCREASE ABOVE THIS

||

MAXIMUM REACTION RATE V_{MAX}

REACTION RATE (V)

SUBSTRATE CONCENTRATION (S)

THE MICHAELIS-MENTEN EQUATION AND THE MICHAELIS CONSTANT

$$v = \frac{V_{max}[S]_0}{[S]_0 + K_m}$$

v : REACTION RATE

$[S]_0$: SUBSTRATE CONCENTRATION BEFORE ENZYME IS ADDED

OHHH...I THINK I'M GETTING A MIGRAINE.

CALM DOWN, KUMI.

YEOW

NOW COMES THE IMPORTANT PART! IN DERIVING THIS COMPLEX EQUATION, MICHAELIS DEFINED A NUMERIC VALUE CALLED THE *MICHAELIS CONSTANT* (K_M), WHICH IS THE SUBSTRATE CONCENTRATION WHEN THE *INITIAL RATE** OF THE REACTION IS HALF OF V_{MAX}.

V_{max}

$\frac{1}{2}V_{max}$

REACTION RATE (V) →

IMPORTANT!

K_m

SUBSTRATE CONCENTRATION (S)

* THE INITIAL RATE IS THE RATE AT THE BEGINNING OF THE REACTION, WHEN THE SUBSTRATE CONCENTRATION AND REACTION SPEED ARE STILL LINEARLY RELATED.

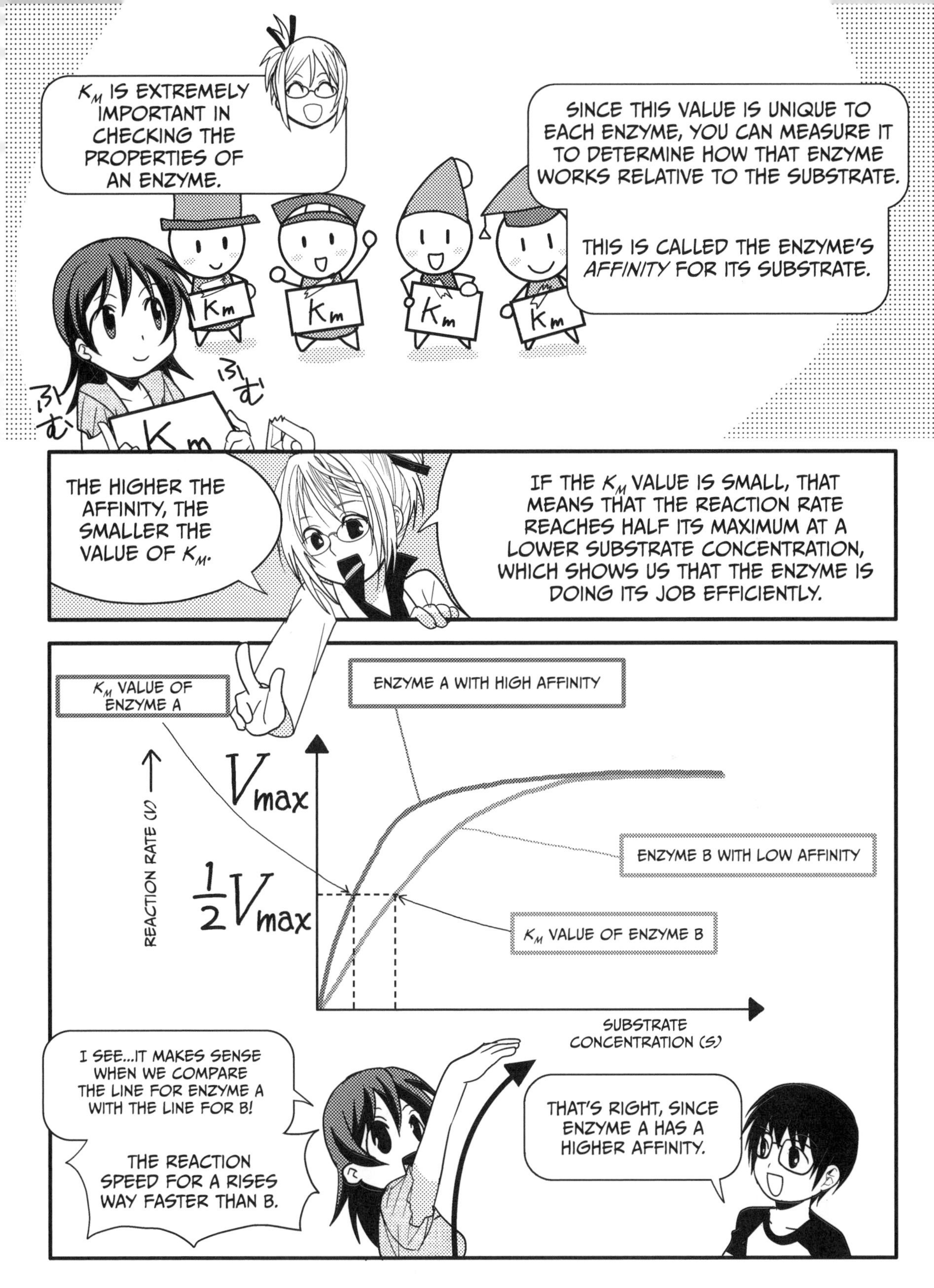
K_M IS EXTREMELY IMPORTANT IN CHECKING THE PROPERTIES OF AN ENZYME.
SINCE THIS VALUE IS UNIQUE TO EACH ENZYME, YOU CAN MEASURE IT TO DETERMINE HOW THAT ENZYME WORKS RELATIVE TO THE SUBSTRATE.
THIS IS CALLED THE ENZYME'S *AFFINITY* FOR ITS SUBSTRATE.
Km
Km
Km
Km
Km
ふむ
ふむ
THE HIGHER THE AFFINITY, THE SMALLER THE VALUE OF K_M.
IF THE K_M VALUE IS SMALL, THAT MEANS THAT THE REACTION RATE REACHES HALF ITS MAXIMUM AT A LOWER SUBSTRATE CONCENTRATION, WHICH SHOWS US THAT THE ENZYME IS DOING ITS JOB EFFICIENTLY.
K_M VALUE OF ENZYME A
ENZYME A WITH HIGH AFFINITY
REACTION RATE (V) →
V_{max}
ENZYME B WITH LOW AFFINITY
$\frac{1}{2}V_{max}$
K_M VALUE OF ENZYME B
SUBSTRATE CONCENTRATION (S)
I SEE...IT MAKES SENSE WHEN WE COMPARE THE LINE FOR ENZYME A WITH THE LINE FOR B!
THE REACTION SPEED FOR A RISES WAY FASTER THAN B.
THAT'S RIGHT, SINCE ENZYME A HAS A HIGHER AFFINITY.

LET'S CALCULATE V_{MAX} AND K_M!

NOW, LET'S TRY TO FIGURE OUT THE V_{MAX} AND K_M VALUES OF A PARTICULAR ENZYME!

WE'LL USE DNA POLYMERASE, WHICH IS AN ENZYME FOR SYNTHESIZING DNA, AS AN EXAMPLE.

NUCLEOTIDES, THE BUILDING BLOCKS OF DNA, WILL BE THE SUBSTRATE IN THIS EXAMPLE.

LET'S SAY WE ADD DNA POLYMERASE TO SIX DIFFERENT SOLUTIONS WITH THE FOLLOWING SUBSTRATE CONCENTRATIONS:*

SUBSTRATE CONCENTRATION
0 μM
1 μM
2 μM
4 μM
10 μM
20 μM

THEN WE LET THE REACTION RUN FOR 60 MINUTES AT A TEMPERATURE OF 37°C.

LET'S ASSUME THIS EXPERIMENT PRODUCED THE FOLLOWING MEASUREMENTS:

SUBSTRATE CONCENTRATION	PRODUCED	REACTION PRODUCT
0 μM	→	0 pmol
1 μM	→	9 pmol
2 μM	→	15 pmol
4 μM	→	22 pmol
10 μM	→	35 pmol
20 μM	→	43 pmol

THESE RESULTS SHOW THE CONCENTRATION OF REACTION PRODUCT FORMED OVER THE COURSE OF AN HOUR, SO WE CAN DIVIDE THESE VALUES BY 1 HOUR TO TURN THEM INTO REACTION RATES.**

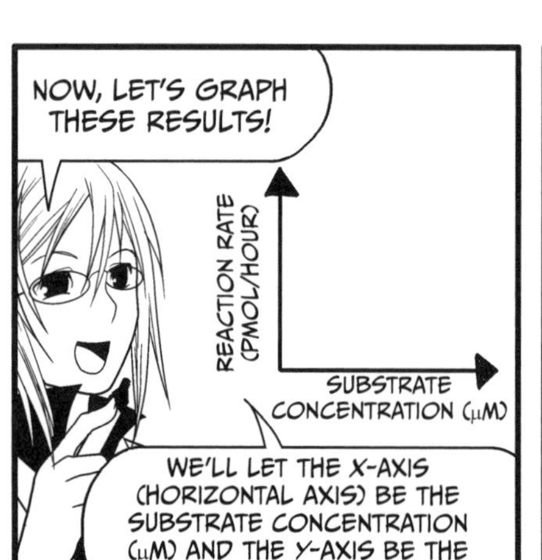

* ACTUALLY, WE'D ALSO NEED TO ADD TEMPLATE DNA, MAGNESIUM IONS, AND OTHER ELEMENTS, BUT LET'S KEEP THINGS SIMPLE!

** FOR EXAMPLE, IF WE START WITH A SUBSTRATE CONCENTRATION OF 1 μM, OUR REACTION RATE IS 9 PMOL PER 1 HOUR, OR, SIMPLY, 9 PMOL/HOUR.

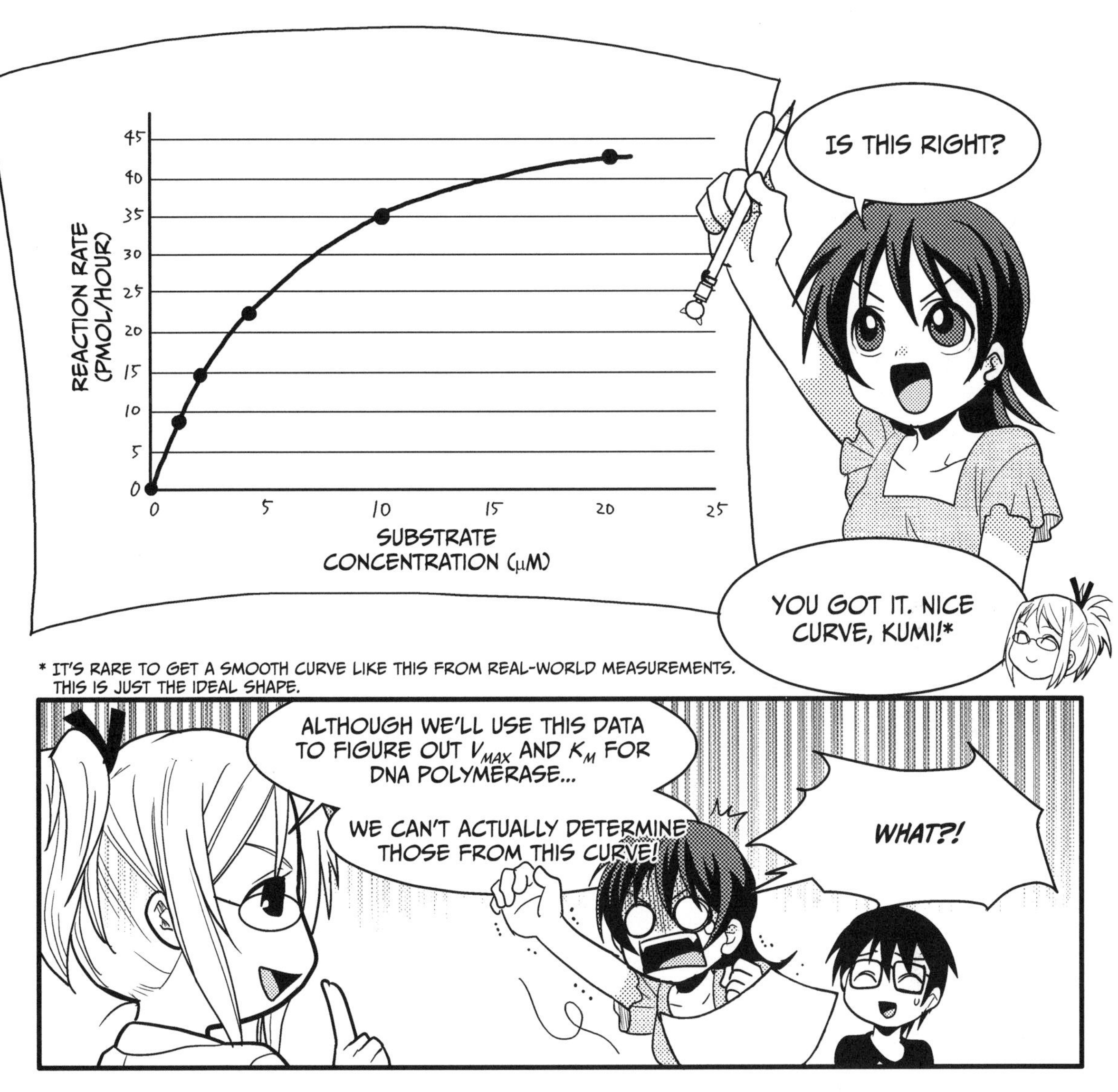

* IT'S RARE TO GET A SMOOTH CURVE LIKE THIS FROM REAL-WORLD MEASUREMENTS. THIS IS JUST THE IDEAL SHAPE.

IT'S EASY! FOR EXAMPLE, THE RECIPROCAL OF 2 IS 1/2 = 0.5.

THE RECIPROCAL OF *A* IS OBTAINED BY CALCULATING *1/A*.

THE RECIPROCAL OF 2 IS 1/2 = 0.5

THE RECIPROCAL OF 3 IS 1/3 = 1.333...

THE RECIPROCAL OF 4 IS 1/4 = 0.25

I'LL EXPLAIN THE THEORY LATER. FOR NOW, LET'S JUST TRY DRAWING A GRAPH!

OKAY!

IF I TAKE THE RECIPROCALS OF THE SUBSTRATE CONCENTRATION VALUES (WHICH ARE ALONG THE HORIZONTAL AXIS), I GET THE FOLLOWING RESULTS:

THE RECIPROCAL OF 1 IS 1

THE RECIPROCAL OF 2 IS 0.5

THE RECIPROCAL OF 4 IS 0.25

THE RECIPROCAL OF 10 IS 0.1

THE RECIPROCAL OF 20 IS 0.05

I CAN'T TAKE THE RECIPROCAL OF ZERO, SO I'M LEAVING IT OUT!

THE VALUES ON THE VERTICAL AXIS (OTHER THAN ZERO) WERE 9, 15, 22, 35, AND 43, SO:

THE RECIPROCAL OF 9 IS 1/9, AND 1 ÷ 9 = 0.111.
THE RECIPROCAL OF 15 IS 1/15, AND 1 ÷ 15 = 0.066.

NOW LET ME JUST WORK THROUGH THE REST...

THE RECIPROCAL OF 9 IS APPROXIMATELY 0.111

THE RECIPROCAL OF 15 IS APPROXIMATELY 0.066

THE RECIPROCAL OF 22 IS APPROXIMATELY 0.045

THE RECIPROCAL OF 35 IS APPROXIMATELY 0.028

THE RECIPROCAL OF 43 IS APPROXIMATELY 0.023

CHECK!

WHY DO WE TAKE RECIPROCALS?

Yes, why *do* we take reciprocals? It's a mystery, isn't it? Heh heh heh...

Don't look at me! Math isn't really my thing.

Don't worry. We'll solve the mystery together. ♪

First, focus on V_{max}. In the following graph, you can see that as the substrate concentration values get larger, the reaction rate measurements approach V_{max}.

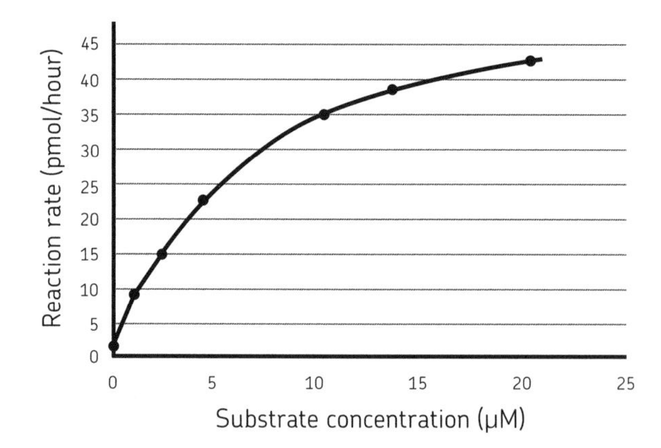

You're right. As the substrate concentration increases, the reaction rate approaches its *limit*, V_{max}, which is the maximum rate.

If we keep taking measurements like this, we'll be able to figure out V_{max}, 1/2 V_{max}, and K_m, right?

In theory, that's true—we could just take a bunch of measurements, and eventually we'd approach V_{max}, but in practice, it's a bit more difficult than that.

Huh? But why?

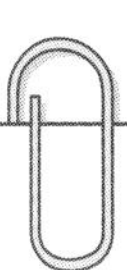

As you get closer to V_{max}, the reaction rate measurements start to bunch up close together.

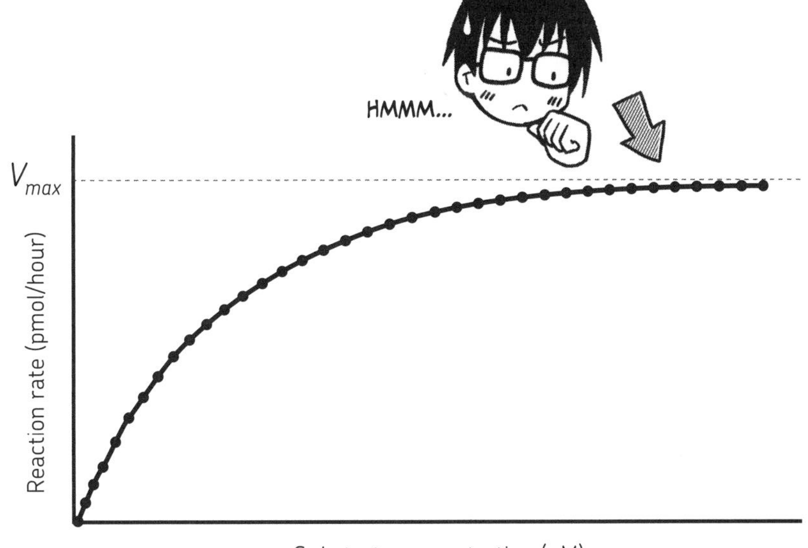

Eventually, the measurements are very close together, and the results just aren't precise enough to show us exactly where the rate maxes out.

Doesn't it seem like the rate just keeps climbing a tiny bit at a time but never gets to V_{max}?

Yeah, it does, but there has to be a way to get around this problem, right?

Of course. We just have to change our way of thinking! For instance, suppose we let the substrate concentration increase more and more, all the way to...infinity!

Hmm...if the substrate concentration were infinite, the measurement result would have to be...V_{max}!

But we can't make calculations using infinity, right? So what can we do?

It's no problem. We'll just use a little trick. We can take advantage of the fact that **the reciprocal of infinity is zero**. In other words, when we take the reciprocal, we know that the *y*-axis value is V_{max} when the *x*-axis is zero!

Although, actually, the numerical value of the *y*-axis at this time is $\frac{1}{V_{max}}$.

It looks a little something like this:

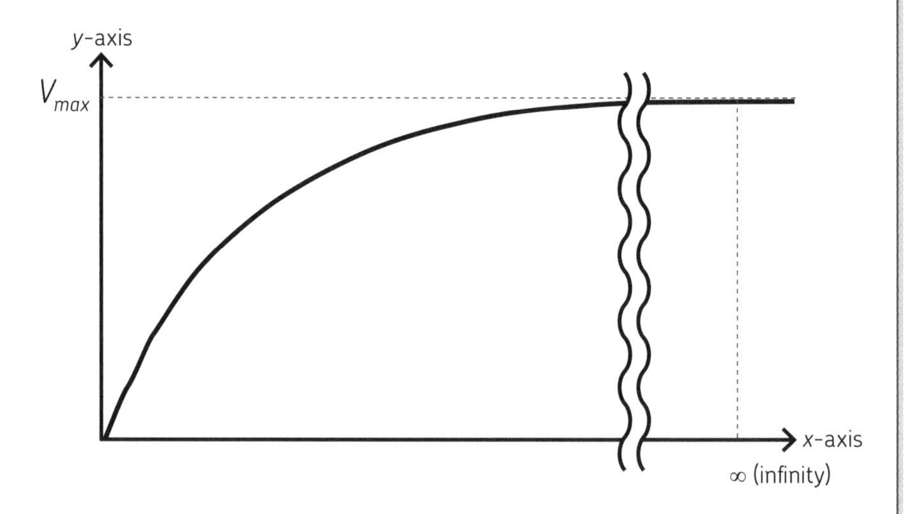

Step 1
When the *x*-axis value is infinity, the *y*-axis value is V_{max}.

Since infinity is too abstract for our calculations, we need to replace it with something that makes sense in the real world, so we just take the reciprocals of the values for both the *x*- and *y*-axes.

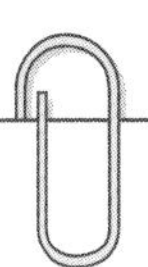

Step 2

When the x-axis value is $\frac{1}{\infty}$, the y-axis value is $\frac{1}{V_{max}}$.

Luckily, we know that the reciprocal of infinity is zero.

Step 3

When the x-axis value is 0, the y-axis value is $\frac{1}{V_{max}}$.

If we create a graph for this, we can get a definite numerical value for V_{max}.

I can't believe it, but...that actually makes sense! Maybe graphs aren't quite as evil as I thought they were...

Now that you understand the theory behind the calculations, don't you feel like your life is richer? Like you understand the universe just a little bit better?

Now let's return to the straight line graph you created earlier and try to obtain the value of V_{max}.

Yeah, let's do it! Graphs and reciprocals are child's play for the likes of me!

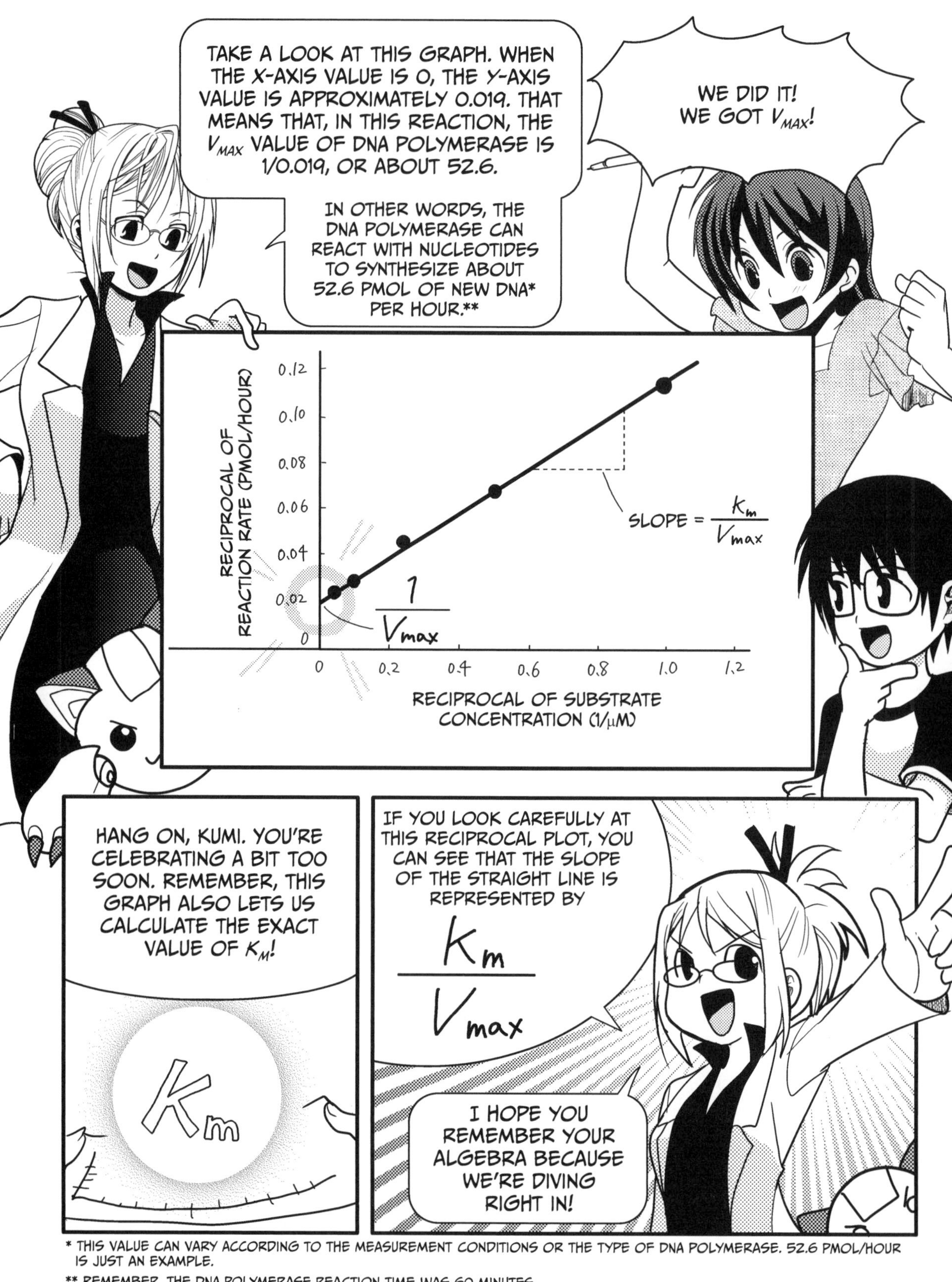

* THIS VALUE CAN VARY ACCORDING TO THE MEASUREMENT CONDITIONS OR THE TYPE OF DNA POLYMERASE. 52.6 PMOL/HOUR IS JUST AN EXAMPLE.

** REMEMBER, THE DNA POLYMERASE REACTION TIME WAS 60 MINUTES.

LET'S USE THE MICHAELIS-MENTEN EQUATION TO FIGURE OUT K_M AND V_{MAX}!

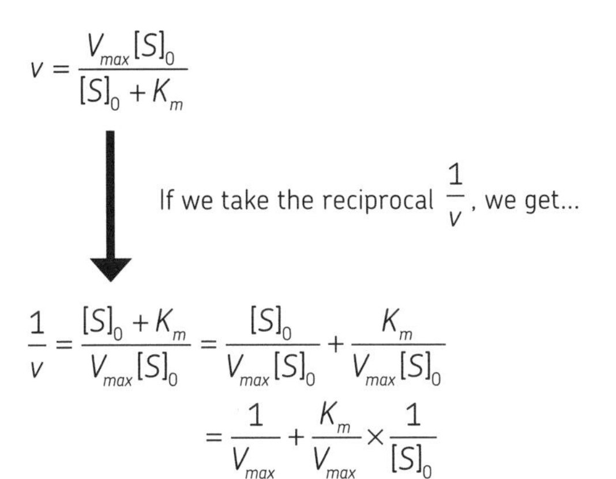

$$v = \frac{V_{max}[S]_0}{[S]_0 + K_m}$$

If we take the reciprocal $\frac{1}{v}$, we get...

$$\frac{1}{v} = \frac{[S]_0 + K_m}{V_{max}[S]_0} = \frac{[S]_0}{V_{max}[S]_0} + \frac{K_m}{V_{max}[S]_0}$$

$$= \frac{1}{V_{max}} + \frac{K_m}{V_{max}} \times \frac{1}{[S]_0}$$

In other words, we can draw the graph of the straight line $y = ax + b$, where

$y : \frac{1}{v}$

$x : \frac{1}{[S]_0}$

$a : \frac{K_m}{V_{max}}$

$b : \frac{1}{V_{max}}$

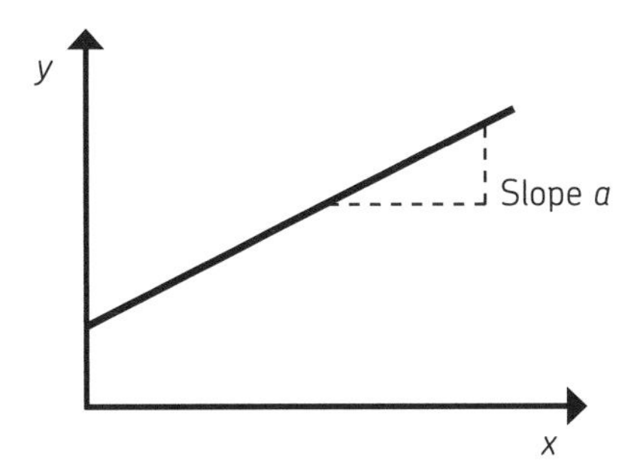

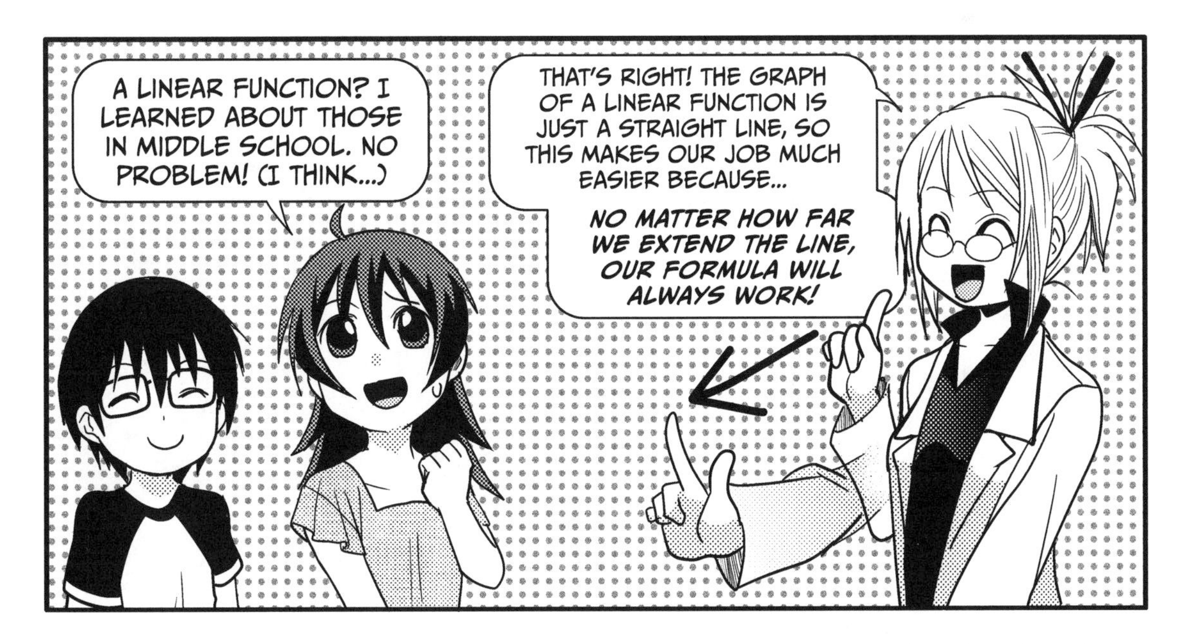

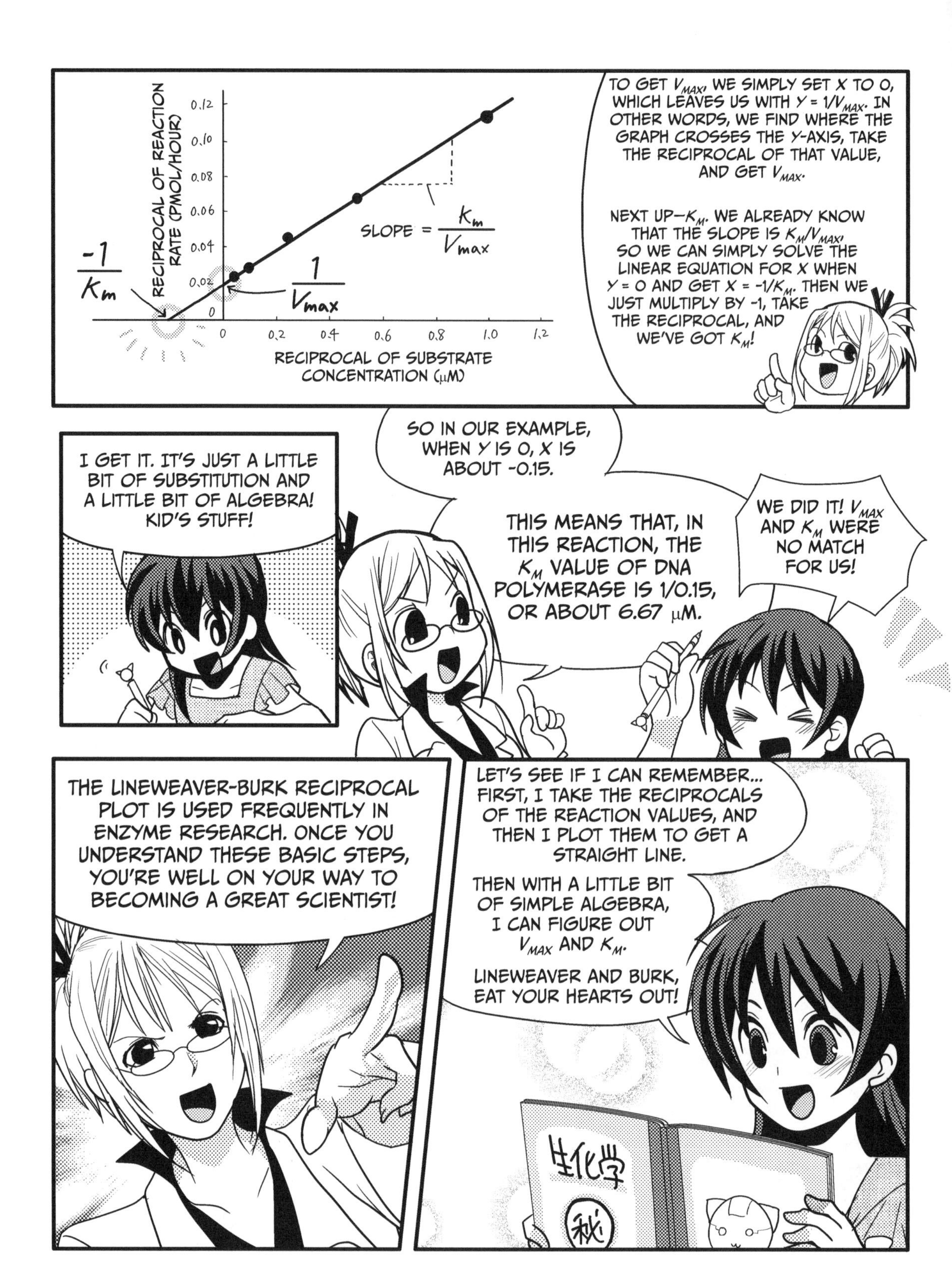

RECIPROCAL OF REACTION RATE (PMOL/HOUR)
0.12
0.10
0.08
0.06
0.04
0.02
0
0
0.2
0.4
0.6
0.8
1.0
1.2
RECIPROCAL OF SUBSTRATE CONCENTRATION (μM)
$\frac{-1}{K_m}$
$\frac{1}{V_{max}}$
SLOPE = $\frac{K_m}{V_{max}}$
TO GET V_{MAX}, WE SIMPLY SET X TO 0, WHICH LEAVES US WITH $Y = 1/V_{MAX}$. IN OTHER WORDS, WE FIND WHERE THE GRAPH CROSSES THE Y-AXIS, TAKE THE RECIPROCAL OF THAT VALUE, AND GET V_{MAX}.
NEXT UP—K_M. WE ALREADY KNOW THAT THE SLOPE IS K_M/V_{MAX}, SO WE CAN SIMPLY SOLVE THE LINEAR EQUATION FOR X WHEN Y = 0 AND GET $X = -1/K_M$. THEN WE JUST MULTIPLY BY -1, TAKE THE RECIPROCAL, AND WE'VE GOT K_M!
I GET IT. IT'S JUST A LITTLE BIT OF SUBSTITUTION AND A LITTLE BIT OF ALGEBRA! KID'S STUFF!
SO IN OUR EXAMPLE, WHEN Y IS 0, X IS ABOUT -0.15.
THIS MEANS THAT, IN THIS REACTION, THE K_M VALUE OF DNA POLYMERASE IS 1/0.15, OR ABOUT 6.67 μM.
WE DID IT! V_{MAX} AND K_M WERE NO MATCH FOR US!
THE LINEWEAVER-BURK RECIPROCAL PLOT IS USED FREQUENTLY IN ENZYME RESEARCH. ONCE YOU UNDERSTAND THESE BASIC STEPS, YOU'RE WELL ON YOUR WAY TO BECOMING A GREAT SCIENTIST!
LET'S SEE IF I CAN REMEMBER... FIRST, I TAKE THE RECIPROCALS OF THE REACTION VALUES, AND THEN I PLOT THEM TO GET A STRAIGHT LINE.
THEN WITH A LITTLE BIT OF SIMPLE ALGEBRA, I CAN FIGURE OUT V_{MAX} AND K_M.
LINEWEAVER AND BURK, EAT YOUR HEARTS OUT!
生化学
秘

4. Enzymes and Inhibitors

Why in the world do we have to use equations and graphs—which are no fun at all—to get the values of V_{max} and K_m?

One reason is to demonstrate that enzyme reactions are extremely precise and predictable. They proceed in strict accordance with chemical and mathematical laws.

More importantly, for scholars doing enzyme research or research concerning the world surrounding enzymes, the values of V_{max} and K_m are very useful.

One of these areas of research deals with the relationship between enzymes and inhibitors. An *inhibitor* is a substance that affects the binding of an enzyme and substrate, or affects the enzyme itself. As a result, it inhibits the enzyme's activity.

Many inhibitors are created artificially and used for enzyme research. Since they inhibit enzymes, they are often toxic to living organisms. However, they can also be used in positive ways—to kill cancer cells, for example.

Inhibitors also exist in the natural world. In that case, they're known as "cellular" enzyme inhibitors rather than "medicinal" inhibitors. For example, some inhibitors are created directly in an organism's cells and play important roles in regulating enzyme reactions. Plants can create inhibitors as well. The seeds of certain legumes, for example, contain cellular enzyme inhibitors such as α-amylase inhibitors or trypsin inhibitors. These are known as *anti-nutritional factors*. Scientists believe that these inhibitors might be part of the plant's natural defense system, which helps protect them against predation.

If the structure of an inhibitor is similar to that of the substrate, it can bond quite easily with the enzyme. However, since the shape of an inhibitor is subtly different from that of the substrate, no reaction will occur, and once the enzyme has already bonded with another substance, it isn't able to bond with its substrate. Actually, there are several different types of enzyme inhibition, and we can tell which is occuring by using the Michaelis-Menten equation.

Two of the types of enzyme inhibition are competitive inhibition and non-competitive inhibition.

Competitive inhibition occurs when a substance that is very similar to the substrate bonds with the enzyme, inhibiting the enzyme reaction.

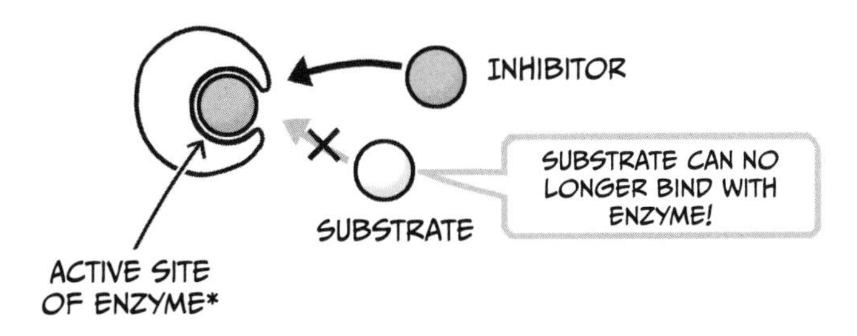

* THE ACTIVE SITE IS THE PART WHERE THE SUBSTRATE BONDS TO THE ENZYME IN ORDER TO UNDERGO A CATALYTIC ACTION.

Since the enzymes remain active, the maximum reaction rate V_{max} is not affected, but the K_m value will rise, since some of the enzymes accidentally bind to inhibitors. With inhibitors getting in the way, it takes longer for enzymes to bind to substrates, producing the same result as if a lower substrate concentration was present.

If a competitive inhibitor is added, the slope of the straight line will be greater on a Lineweaver-Burk reciprocal plot, as shown below. Note that the intersection with the y-axis, $1/V_{max}$, doesn't change.

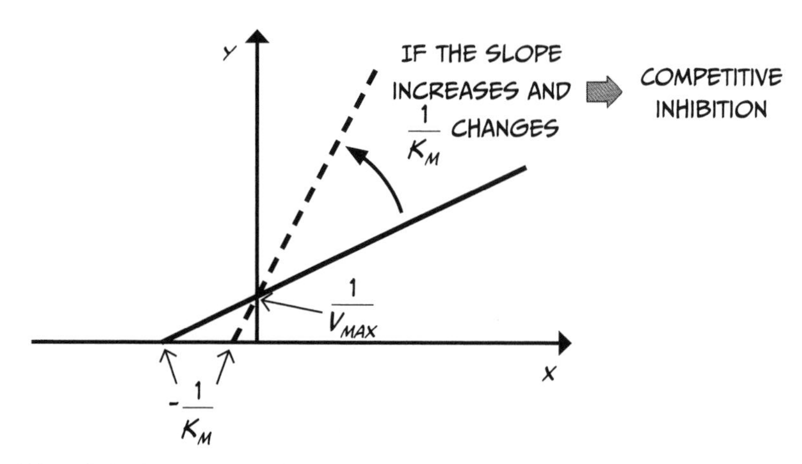

Although the intersection with the y-axis doesn't change, the intersection with the x-axis does. This means that if you add an unknown inhibitor to an enzyme reaction, take measurements, graph the result, and see that $1/K_m$ has increased, you know it's a competitive inhibitor.

Non-competitive inhibition is when an inhibitor bonds at a part of the enzyme unrelated to the substrate bonding site to inhibit the enzyme reaction.

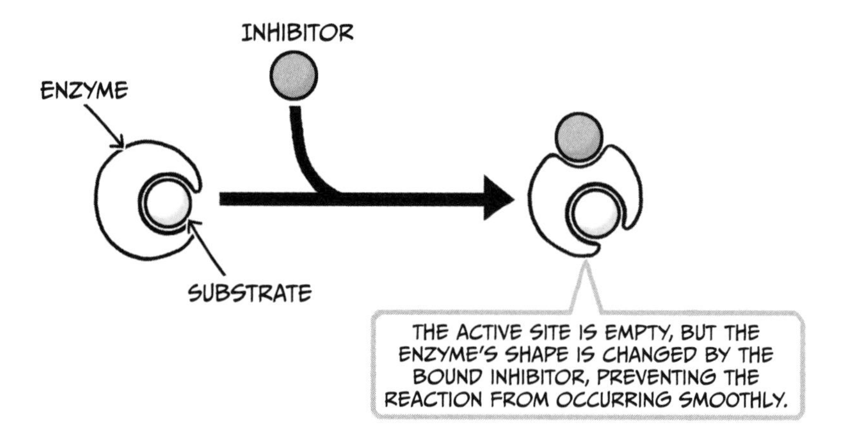

In this case, the inhibitor doesn't block the substrate—it takes the enzymes out of the game. K_m is unaffected because it depends on substrate concentration. Instead, the maximum reaction rate will steadily decrease as the amount of inhibitor increases. The same thing would happen if the enzyme concentration decreased.

This means that when a non-competitive inhibitor is added, the slope of the straight line will again be greater on a Lineweaver-Burk reciprocal plot, as shown below. But unlike competitive inhibition, the intersection with the *y*-axis will change, while the intersection with the *x*-axis will stay the same.

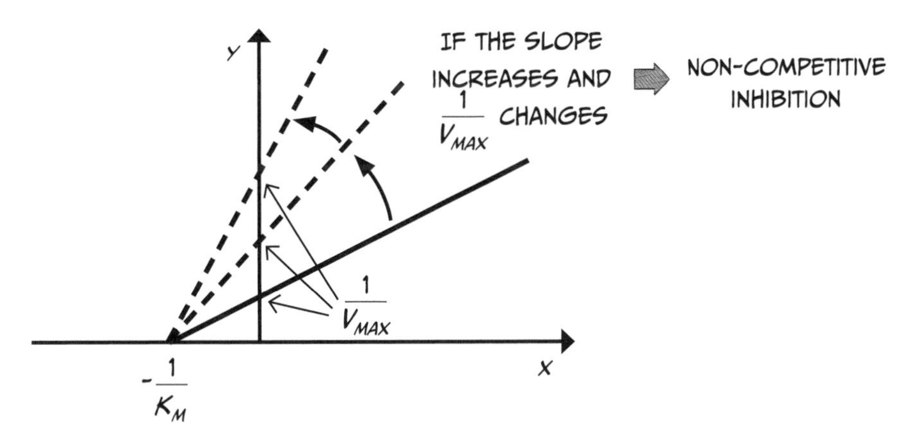

Since inhibitors significantly affect enzyme reactions, they are often used in research for elucidating enzyme structures or reaction mechanisms. They are also used in cancer research by applying small amounts to inhibit enzymes that benefit the cancer cells.

ALLOSTERIC ENZYMES

CHECK!

Although we've introduced enzyme reactions for which the Michaelis-Menten equation applies, there are also many enzymes that exhibit activity for which the Michaelis-Menten equation does not apply.

One of these is a type of enzyme called an *allosteric enzyme*, which can respond to environmental signals with changes in activity. This is referred to as an *allosteric effect*. For example, a substrate that bonds to one subunit may bring about a change in the three-dimensional structure of the enzyme, which causes the substrate to bond more easily with another subunit.

In a case like that, the curve representing the relationship between the substrate concentration and the reaction speed will be an S-shaped *sigmoid curve* rather than the hyperbola that is typical of the Michaelis-Menten equation.

We'll skip the details here, but not all enzyme reactions are simple ones that occur according to certain fixed equations—in fact, in the body, most are extremely complex and diverse.

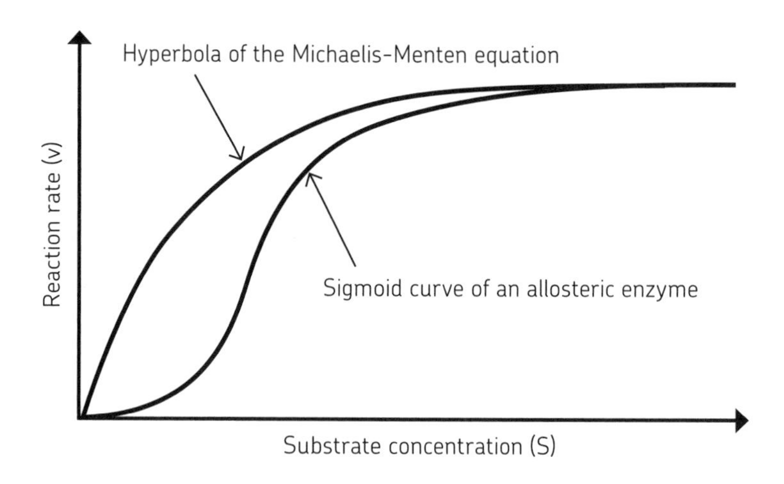

TODAY'S LESSON WAS PRETTY TOUGH. HOW ARE YOU FEELING?
SWISH
SWISH
STRETCH
GOOD!
THE CALCULATIONS STARTED TO GET KIND OF TRICKY TOWARD THE END, BUT I'M HAPPY THAT I MANAGED TO STICK IT OUT!
PLUS, I FOUND OUT THAT ENZYMES ARE REALLY NICE LITTLE GUYS!
HUH?

THEY LOWER DIFFICULT-TO-CLIMB WALLS AND GIVE US THE BOOST WE NEED TO HOP OVER!
NO PROBLEMO
THANKS!

WELL, IT DOESN'T QUITE WORK THAT WAY IN REAL LIFE, OF COURSE.

BUT NOW THAT YOU MENTION IT, I AGREE THAT ENZYMES ARE HELPFUL LITTLE GUYS.
HEH
HEH
HEH
WHEN THERE'S AN IMPENETRABLE BARRIER BETWEEN A CERTAIN BOY AND THE GIRL HE LIKES...
...THAT BARRIER CAN BE BROUGHT DOWN WITH THE HELP OF A KINDLY ENZYME!
THIS ENZYME'S ROLE IS TO ACCELERATE THE RELATIONSHIP BETWEEN THESE TWO PEOPLE.
SNICKER
HEY, PROFESSOR KUROSAKA! MAYBE WE SHOULD TALK ABOUT THE NEXT LESSON!
WHAT THE...?

SIGH...I SUPPOSE SO. THE NEXT LESSON WILL BE THE LAST, SO YOU BETTER PREPARE YOURSELVES! *ESPECIALLY YOU, NEMOTO.*
THE LAST LESSON? THAT'S A LITTLE SAD...
HUFF
PUFF

MOLECULAR BIOLOGY AND THE BIOCHEMISTRY OF NUCLEIC ACIDS

MORNING! BREAKFAST TIME, I HOPE?
SURE IS. I'M MAKING IT NOW.
I'M STARVING!
IT'S GOOD TO SEE YOU EATING BREAKFAST AGAIN, KUMI.
SCARF
OM NOM
YOUR FATHER'S RIGHT! I HOPE YOU'RE DONE WITH ALL THAT FASTING NONSENSE. A GROWING GIRL NEEDS A GOOD BREAKFAST!

WELL, NOW I KNOW BETTER. I'VE GOT TO EAT SO MY BODY CAN CREATE ATP!
I'M THROUGH WITH STUPID FAD DIETS.

?
A-T-P?

BESIDES, AFTER THE FINAL LESSON TODAY...
FOR ALL YOUR HARD WORK, YOU'VE EARNED THESE— MY FOOLPROOF, SCIENTIFIC DIETING SECRETS!
THANK YOU, PROFESSOR!
KUMI'S IMAGINATION
...I'LL KNOW ALL THE SECRETS OF A HEALTHY DIET!

DOWN THE HATCH!
HA HA HA HA
BWA HA HA
OM NOM NOM

OKAY, LET'S GET STARTED!
YEAH!
KREBS UNIVERSITY

1. What Is Nucleic Acid?

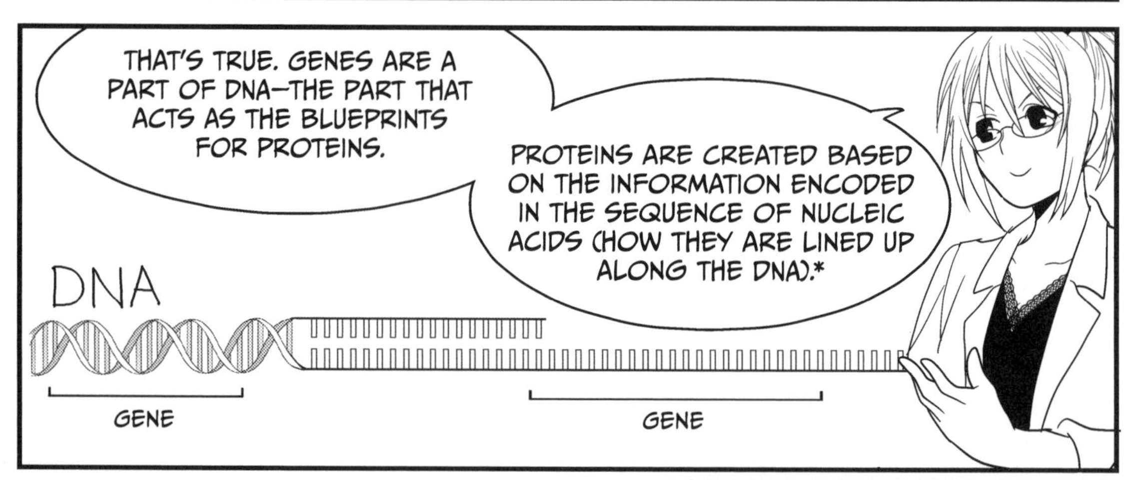

* GENES ALSO CONTAIN THE "BLUEPRINTS" FOR RNA.

RECIPE
PROTEIN
PROTEIN
(SEE PAGE 19.)
AH, I REMEMBER! GENES ARE WRITTEN IN THE DNA INSIDE THE NUCLEUS!
AND PROTEINS ARE CREATED BASED ON THOSE BLUEPRINTS.
MM HM!
THAT'S RIGHT! NOW WE'LL TALK ABOUT THAT PROCESS IN MORE DETAIL.
THERE ARE ACTUALLY TWO TYPES OF NUCLEIC ACID: DNA (DEOXYRIBONUCLEIC ACID) AND RNA (RIBONUCLEIC ACID).
DNA
RNA

RNA AND DNA ARE BOTH NECESSARY FOR THE CREATION OF PROTEINS.
TRANSCRIPTION
TRANSLATION
DNA
(CONTAINS GENES)
RNA
PROTEIN
I ALREADY KNOW A LITTLE BIT ABOUT DNA...
...BUT RNA IS A TOTAL MYSTERY TO ME!
DON'T WORRY! RNA ISN'T QUITE AS WELL KNOWN AS DNA...
BUT IT WAS RECENTLY FOUND TO BE AN EXTREMELY IMPORTANT SUBSTANCE TO LIVING ORGANISMS LIKE US.

THE DISCOVERY OF NUCLEIN

The Swiss biochemist Friedrich Miescher (1844–1895) successfully isolated a new substance from a sample of white blood cells. He extracted the cells from used bandages that he received from a nearby hospital.

Friedrich Miescher

White blood cells adhering to used bandages? Ugh, so we must be talking about...

That's right! We're talking about **pus**!

He studied pus?! Gross! I guess biochemistry isn't all glamour and diets...

To extract white blood cells, Miescher processed the pus by adding an enzyme called *protease*, which breaks down protein, and then he used a method called *ether extraction* to remove the lipids.

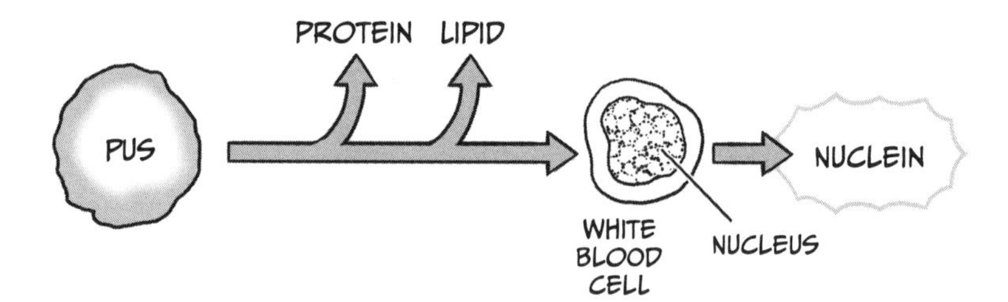

The substance that was obtained from the white blood cells showed strong acidity. Miescher named it *nuclein* since it was found inside the **nucleus** of the white blood cells. Miescher also succeeded in extracting nuclein from salmon sperm. Nuclein later became known as *nucleic acid*, and in the first half of the 20th century, scientists determined that there two types of nucleic acid: DNA and RNA.

NUCLEIC ACID AND NUCLEOTIDES
NOW LET'S EXAMINE THE STRUCTURE OF NUCLEIC ACID!
DNA
RNA
NUCLEIC ACID CONSISTS OF A LONG CHAIN OF SUBSTANCES CALLED *NUCLEOTIDES** CONNECTED TOGETHER.
I SEE.
* NUCLEOTIDES ALSO APPEARED WHEN WE DISCUSSED GRAPHS IN CHAPTER 4.
A NUCLEOTIDE IS FORMED BY BONDING A *BASE*, A *PHOSPHATE*, AND A *PENTOSE* (FIVE-CARBON SUGAR), AS FOLLOWS:
BASE
P
PHOSPHATE
PENTOSE
RIBOSE
OR
DEOXYRIBOSE
NUCLEOTIDE
1 TO 3 PHOSPHATES
RIBONUCLEOTIDE
DEOXYRIBONUCLEOTIDE
UHH...
AND IF JUST A BASE AND A PENTOSE BOND (WITHOUT A PHOSPHATE), THE RESULT IS CALLED A *NUCLEOSIDE*.

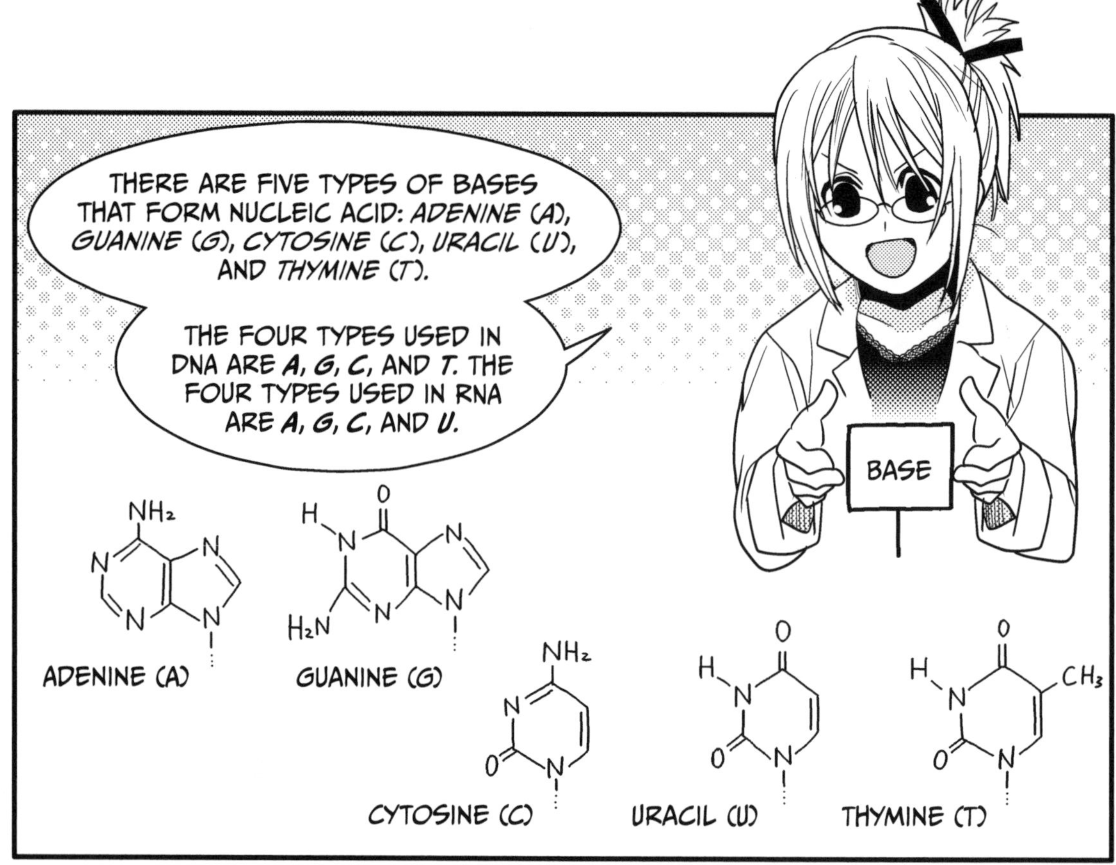
THERE ARE FIVE TYPES OF BASES THAT FORM NUCLEIC ACID: ADENINE (A), GUANINE (G), CYTOSINE (C), URACIL (U), AND THYMINE (T).
THE FOUR TYPES USED IN DNA ARE A, G, C, AND T. THE FOUR TYPES USED IN RNA ARE A, G, C, AND U.
BASE
ADENINE (A)
GUANINE (G)
CYTOSINE (C)
URACIL (U)
THYMINE (T)

A AND G ARE CALLED PURINE BASES, WHILE C, U, AND T ARE CALLED PYRIMIDINE BASES.
I CAN REMEMBER THAT BECAUSE APPLE AND GINGER ARE THE BEST, MOST PURELY DELICIOUS FLAN FLAVORS!
PURINE HAS ONE HEXAGONAL RING AND ONE PENTAGONAL RING.
PURINE
PYRIMIDINE HAS ONLY ONE HEXAGONAL RING.
PYRIMIDINE

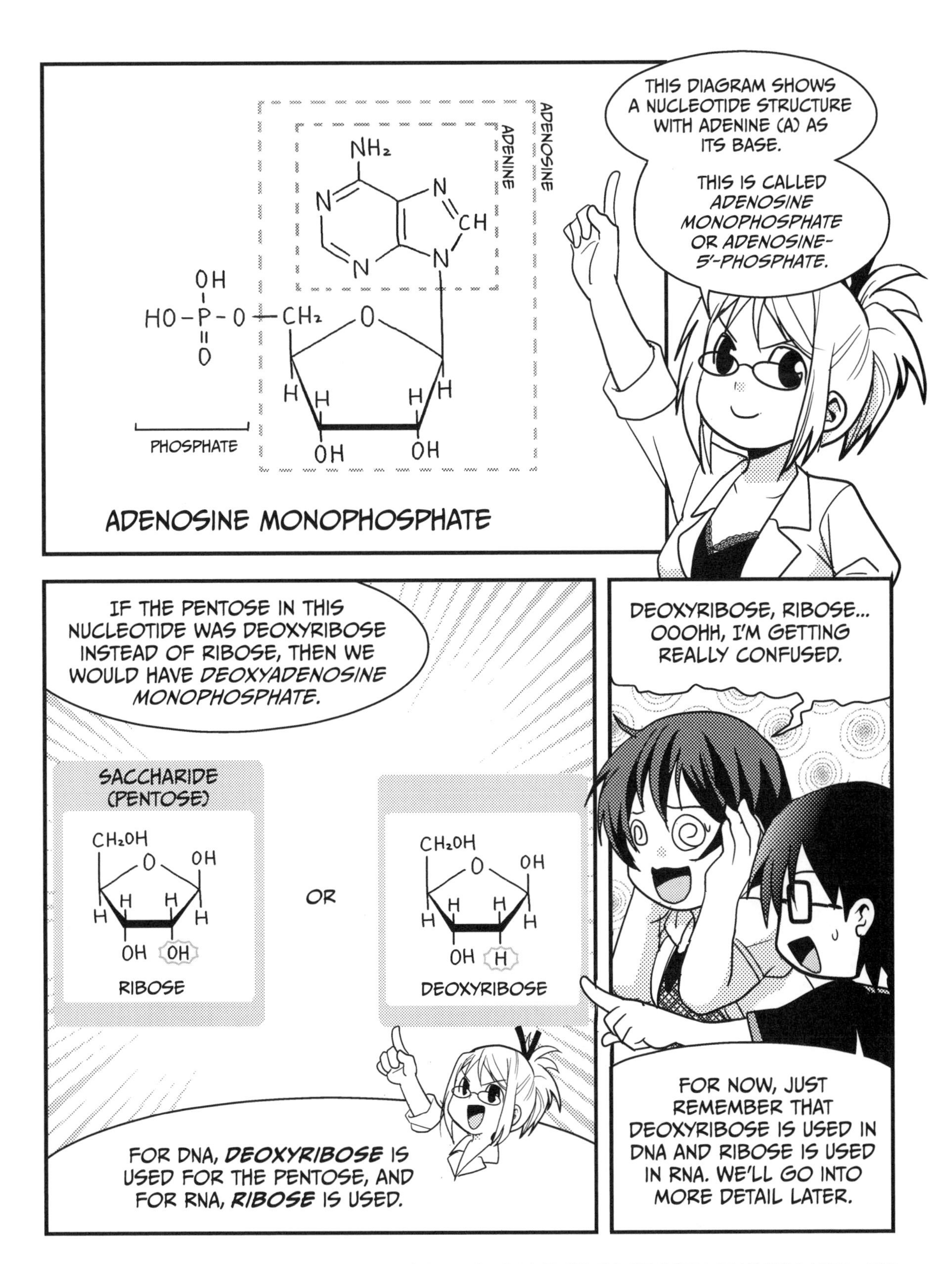
THIS DIAGRAM SHOWS A NUCLEOTIDE STRUCTURE WITH ADENINE (A) AS ITS BASE.
THIS IS CALLED *ADENOSINE MONOPHOSPHATE* OR *ADENOSINE-5′-PHOSPHATE.*
ADENOSINE
ADENINE
NH_2
N
N
CH
N
N
OH
HO–P–O–CH_2
O
O
H
H
H
H
OH
OH
PHOSPHATE
ADENOSINE MONOPHOSPHATE
IF THE PENTOSE IN THIS NUCLEOTIDE WAS DEOXYRIBOSE INSTEAD OF RIBOSE, THEN WE WOULD HAVE *DEOXYADENOSINE MONOPHOSPHATE.*
SACCHARIDE (PENTOSE)
CH_2OH
O
OH
H
H
H
H
OH
OH
RIBOSE
OR
CH_2OH
O
OH
H
H
H
H
OH
H
DEOXYRIBOSE
FOR DNA, ***DEOXYRIBOSE*** IS USED FOR THE PENTOSE, AND FOR RNA, ***RIBOSE*** IS USED.
DEOXYRIBOSE, RIBOSE... OOOHH, I'M GETTING REALLY CONFUSED.
FOR NOW, JUST REMEMBER THAT DEOXYRIBOSE IS USED IN DNA AND RIBOSE IS USED IN RNA. WE'LL GO INTO MORE DETAIL LATER.

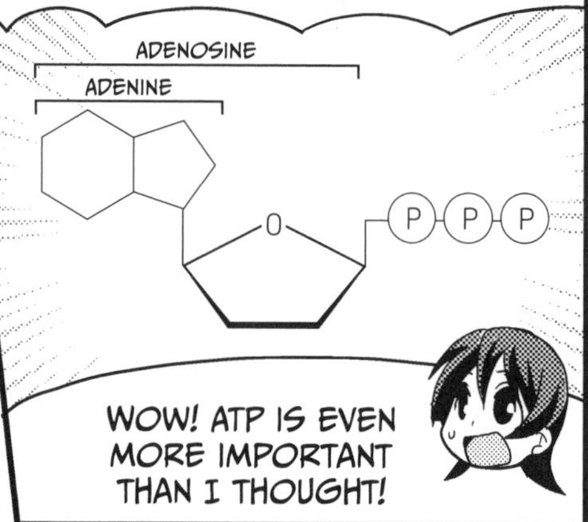

- If the base is guanine (G), we get *guanosine monophosphate* and *deoxyguanosine monophosphate*, for RNA and DNA respectively.
- If it's cytosine (C), we get *cytidine monophosphate* and *deoxycytidine monophosphate*.
- If it's thymine (T), we get *deoxythymidine monophosphate*.*
- If it's uracil (U), we get *uridine monophosphate* and *deoxyuridine monophosphate*.**

* Since the deoxy form of thymine is the most prevalent, the *deoxy* prefix is often left out.

** Although uracil is not a base that's normally in DNA, the deoxy type does exists.

I SEE! THERE ARE TWO TYPES OF PENTOSES AND FIVE TYPES OF BASES. IF THE BASE CHANGES, THE NUCLEOTIDE ALSO CHANGES, RIGHT? IT'S THE SAME WITH UDON AND SOBA!

CURRY

DEPENDING ON WHAT WE COMBINE IT WITH, IT CAN TURN INTO SOBA AND EGG, TEMPURA SOBA, AND SO ON...

SLURP

DON'T FORGET THE GREEN ONIONS AND SEVEN-SPICE POWDER!

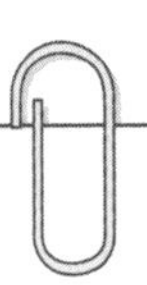

BASE COMPLEMENTARITY AND DNA STRUCTURE

Nucleic acid consists of nucleotides connected in a long, linear chain. The carbons at the 3′ position and the 5′ position of each pentose link to a phosphate, forming a bridge between individual nucleotides. This is called a *polynucleotide*.

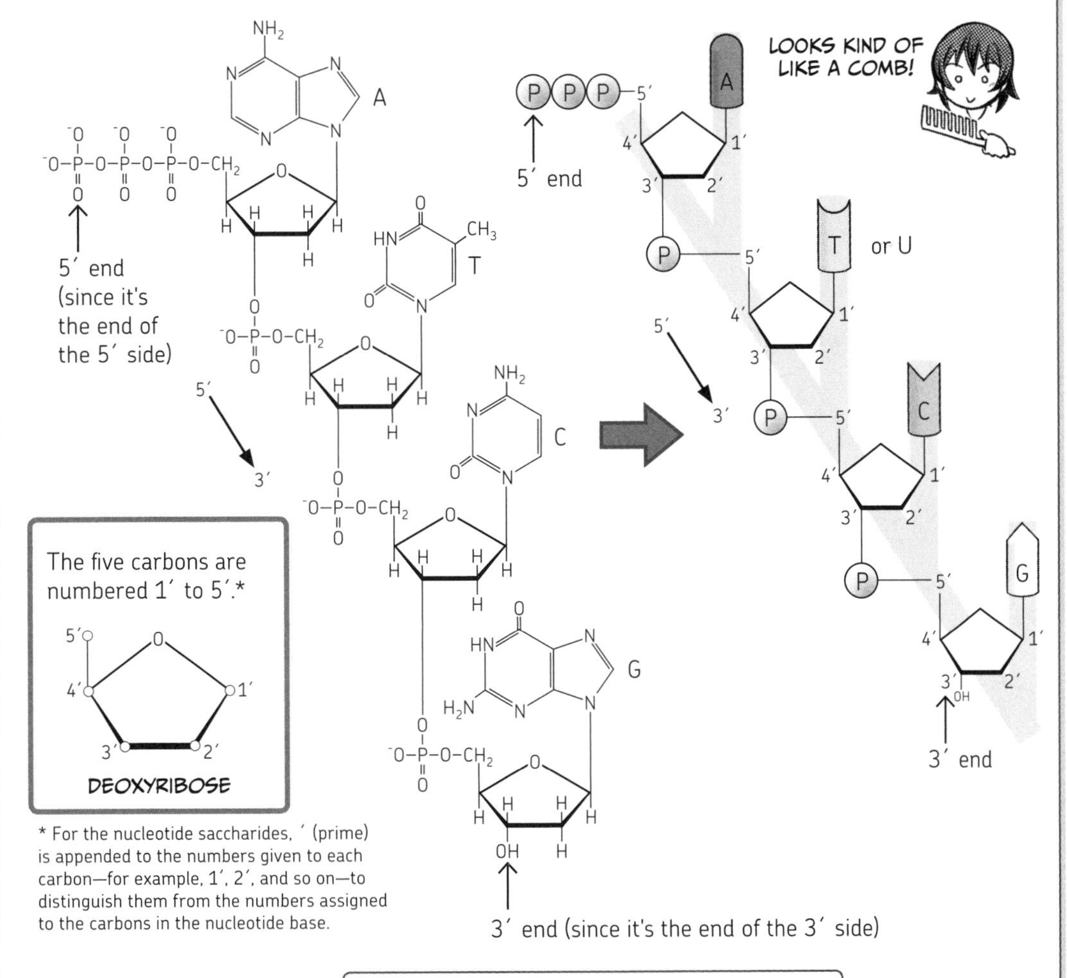

* For the nucleotide saccharides, ′ (prime) is appended to the numbers given to each carbon—for example, 1′, 2′, and so on—to distinguish them from the numbers assigned to the carbons in the nucleotide base.

POLYNUCLEOTIDE

The bases form shapes that protrude like the teeth of a comb on one side of the polynucleotide. For DNA, two polynucleotide strands form a double strand by connecting these bases, which creates a helix structure.

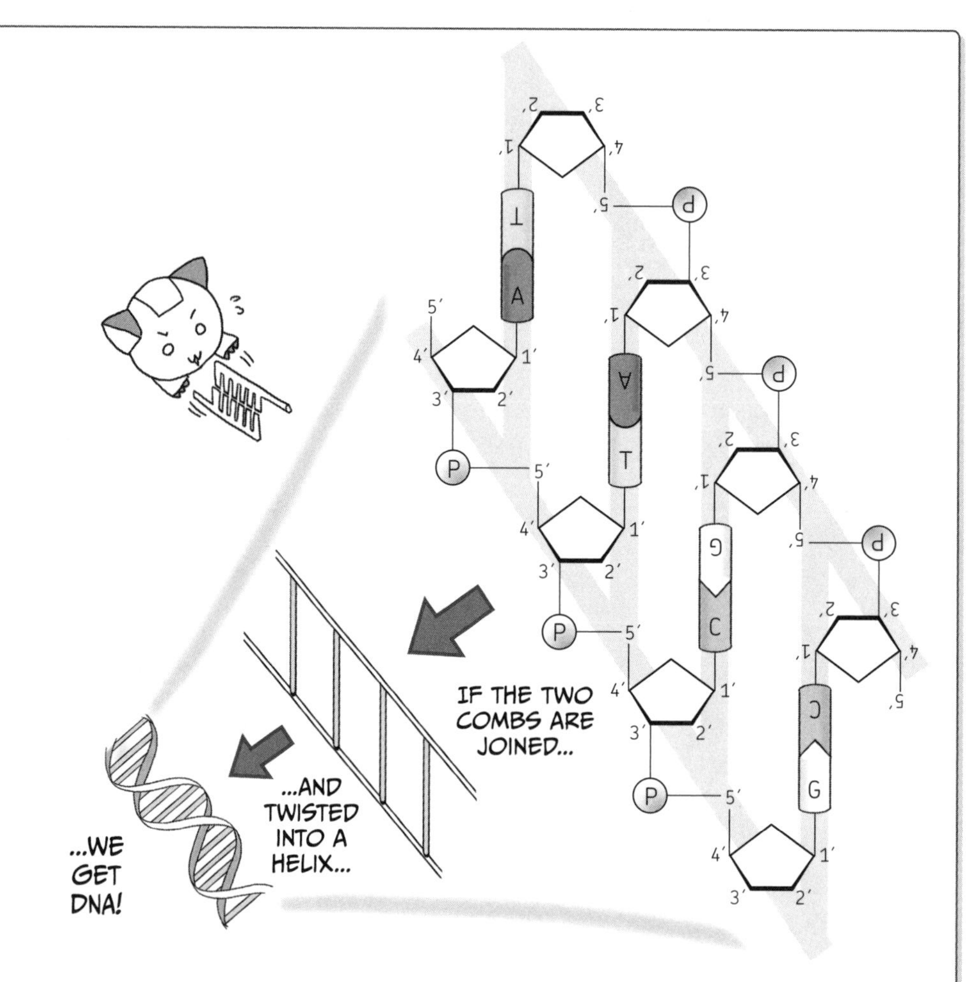

Oh! Those comb-teeth-looking parts stick together and become the rungs in the middle!

Right. In other words, a "pair" is created via hydrogen bonds between two bases, so that the polynucleotides form a double strand. The pairings are always faithful: A pairs with T (with two hydrogen bonds), and G pairs with C (with three hydrogen bonds).

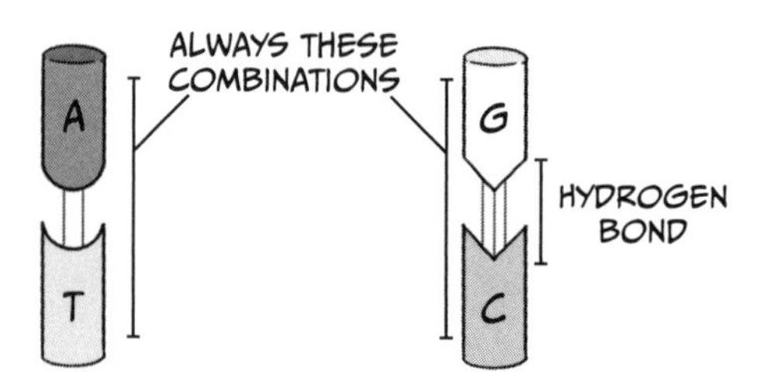

This pairing configuration is called *complementarity*. Because of this complementarity, DNA can be *replicated*—in other words, the two-stranded polynucleotide chain can be separated into individual strands, and those strands can be used as *templates* to build new strands. The result of this replication is two DNA molecules that have the same arrangement of bases.

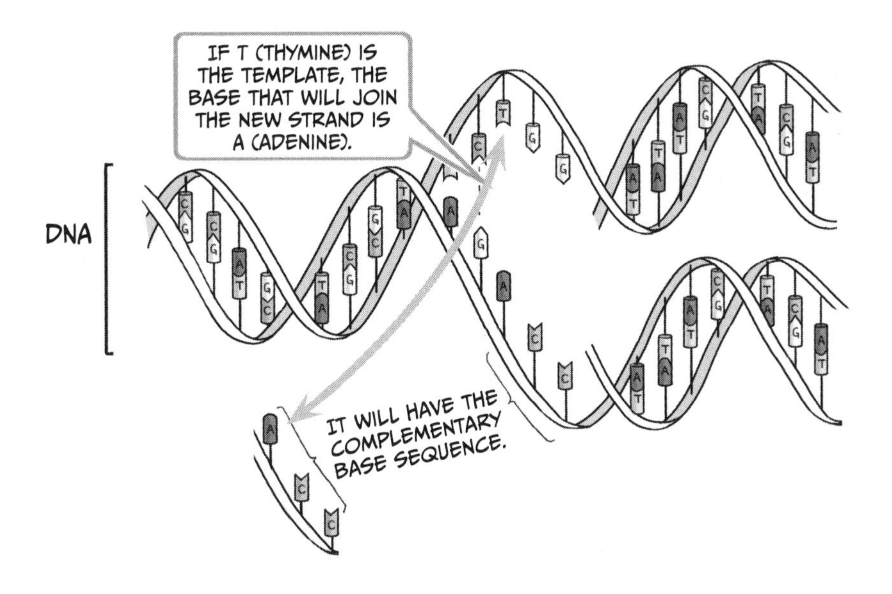

A template is like the mold that you pour batter into to make cupcakes.

I get it! If you have a mold, they're easy to replicate.

If you know what the sequence of the template strand is, you'll know the sequence of a new strand that uses that tempate as a mold!

That's right. The biggest difference between DNA and other biopolymers (proteins, saccharides, and lipids) is that ability to replicate easily.

DNA REPLICATION AND THE ENZYME DNA POLYMERASE

So why is DNA replicated? Since DNA contains our genes, it must be replicated whenever cells divide and multiply.

Genes are passed on from parents to children, as well as from cell to cell. When a cell divides, the same genes must be inherited by the two new cells, so the DNA must be replicated before the cell splits.

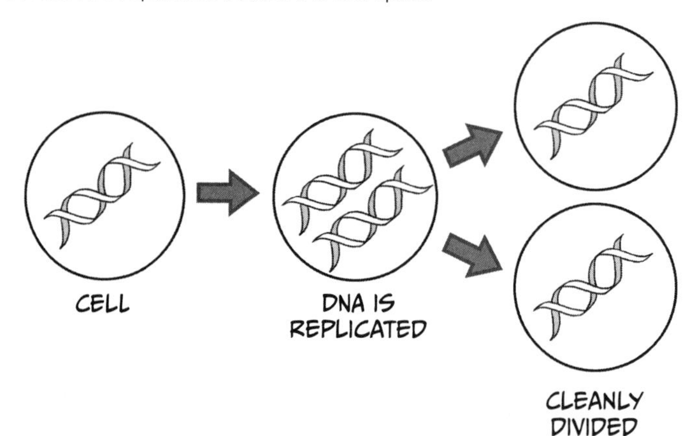

That's...totally cool!

The first person to isolate an enzyme involved in DNA replication—called *DNA polymerase I*—was the American biochemist Arthur Kornberg (1918–2007).

Arthur Kornberg

DNA polymerase was the enzyme we used when we obtained V_{max} and K_m. Do you remember?

Yeah! That's the enzyme that replicates DNA? Enzymes really do a lot, don't they?

In 1956, Kornberg extracted an active enzyme used for synthesizing DNA from a liquid in which *E. coli* was pulverized. This enzyme could be used to artificially synthesize DNA in a test tube, which created a great sensation in the scientific world at the time.

That discovery was seriously important.

It sure was! In fact, Kornberg was awarded the Nobel Prize in Physiology or Medicine in 1959.

DNA polymerase is officially called "DNA-dependent DNA polymerase." This means that a DNA polymerase enzyme requires the presence of an existing DNA template molecule in order to create a new DNA molecule.

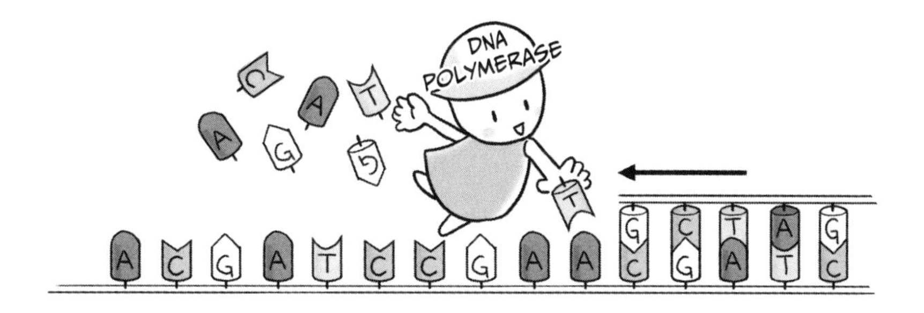

I get it. Since the pairs are complementary, DNA polymerase steadily creates the chain by referring to the template.

The chemical reaction that DNA polymerase catalyzes is shown below.

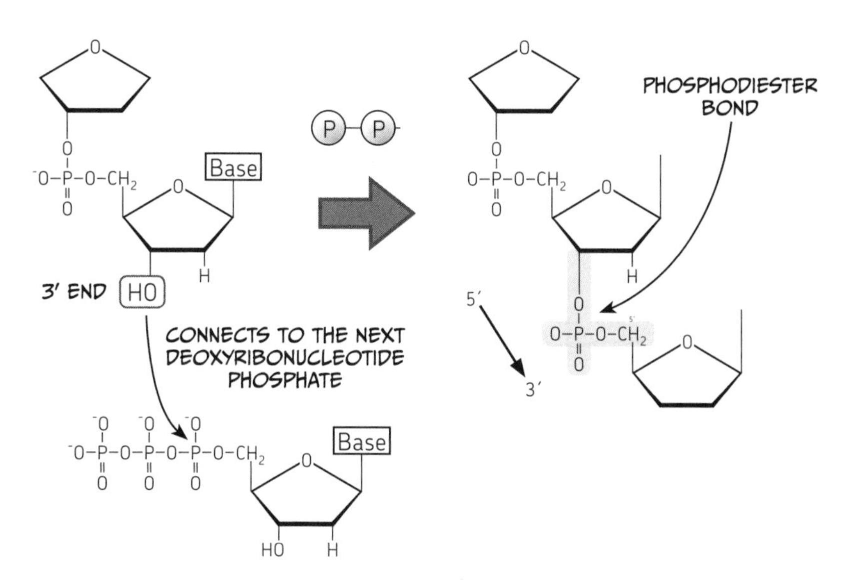

IN THIS REACTION, A NUCLEOTIDE WITH THREE PHOSPHATES ATTACHED IS USED AS THE SUBSTRATE.

In other words, DNA polymerase catalyzes the attachment of the next deoxyribonucleotide phosphate to the 3′ end of a new DNA strand. This attachment is called a *phosphodiester bond*.

The nucleic acids that belong together are joined firmly by the "matchmaker" DNA polymerase.

That's right! Ha ha ha...

...

RNA STRUCTURE

On the other hand, RNA usually exists as a single strand. The enzyme RNA polymerase actually uses DNA as a template when it binds ribonucleotides together into a new RNA strand.

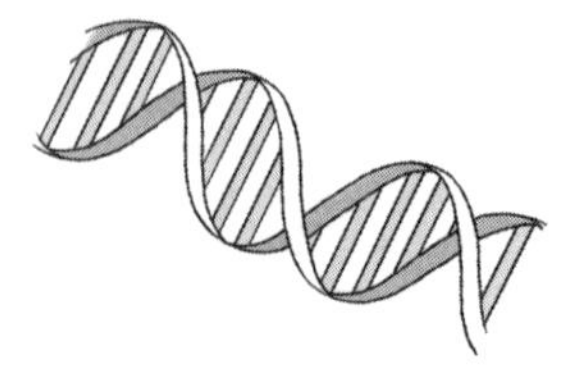

DNA

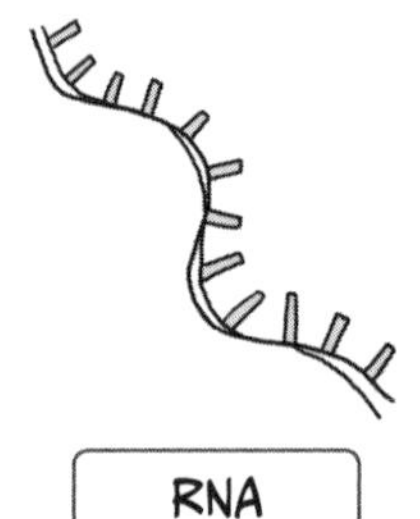

RNA

Oh, that reminds me...earlier you said that RNA is an extremely important substance, but you never said why!

Just hold your horses, Kumi. We'll get into that soon (on page 220, to be exact).

Got it!

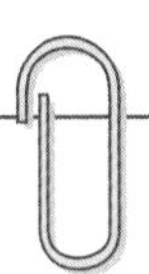

Unlike DNA, RNA usually breaks down soon after it's produced. Kumi, do you remember what we said about deoxyribose?

It's the pentose that's in DNA but not RNA, right?

That's right! In RNA, a hydroxyl group (2-OH) is attached to the carbon at the 2′ position on each ribose, as shown in the following figure. In DNA, only a hydrogen is attached, hence the name *deoxy*ribose.

This hydroxyl group in RNA is actually a real troublemaker. The H could even be considered a biochemical playboy.

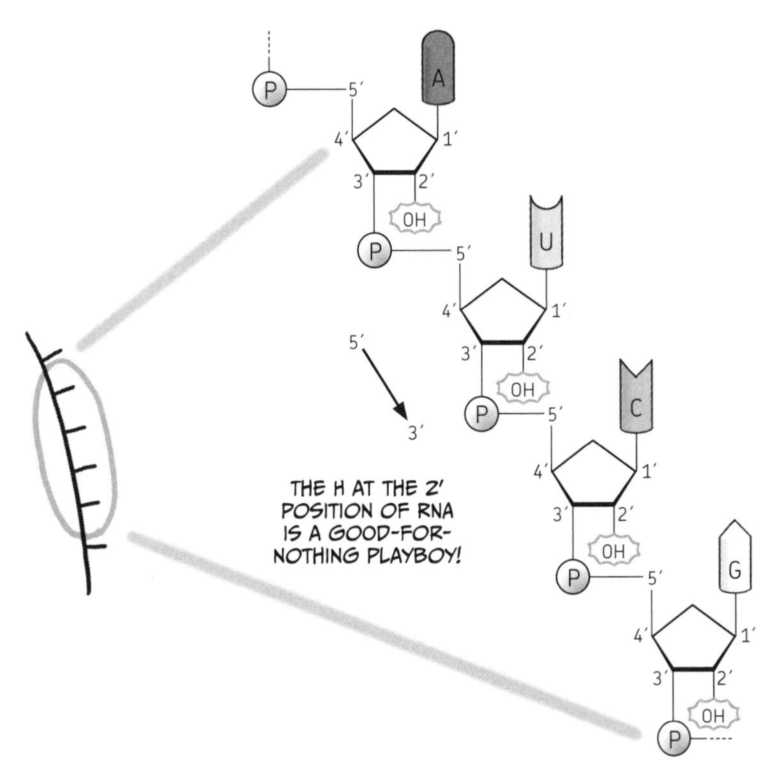

Huh?! What a jerk!

(Of course, ***I*** would always stay faithful...)

Because of the existence of this hydroxyl group, RNA is known to undergo *autolysis* (a self-inflicted decomposition) due to a phenomenon known as *base catalysis*.

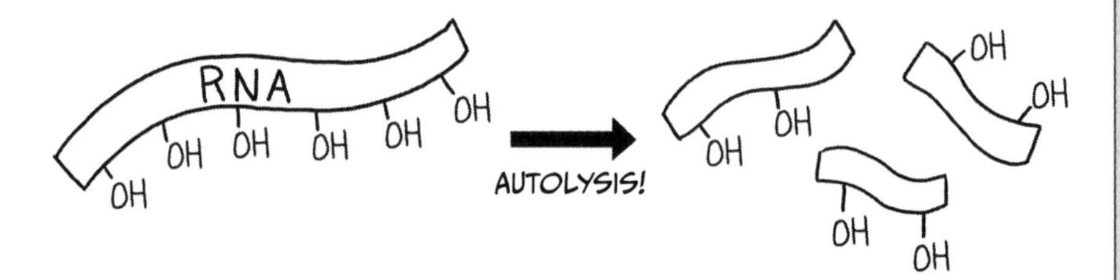

If you think of the O and the H of a hydroxyl group as a couple, then H is the "playboy" that can be tempted by bases that exist around the RNA.

It's important to note that these aren't the A, U, C, and G bases we've been talking about so far. In this context, a *base* is a substance that can accept a proton (H^+), such as a hydroxide ion (OH^-). Bases can be thought of as the chemical opposite of *acids*, because they neutralize each other.

So the 2′-OH proton is extracted by a base?

That's right. Then the lone oxygen goes looking for a new partner to bond with.

Actually, after the proton is extracted, the oxygen (O^-), which carries a negative charge, will bond with the phosphate (P) in the phosphodiester bond at the neighboring 3′ position (that is, the important bond connecting two ribonucleotides).

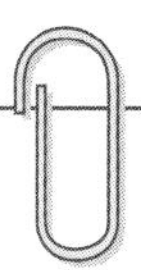

As a result, the phosphodiester bond breaks apart, dividing the RNA chain.

It's all a little sordid, isn't it?

PHOSPHODIESTER BOND

BASE

:OH⁻

ATTACK!

WHAT?!

DIVIDED INTO PIECES

HEY!

Since RNA frequently undergoes this kind of autolysis, it's considered "unstable" and is not very suitable for storing genetic information. DNA is far more stable than RNA since it does not contain the 2′-OH. Therefore, trustworthy DNA is the best for this purpose. A different but still very important role is assigned to RNA instead.

We'll learn about that role next!

2. Nucleic Acid and Genes

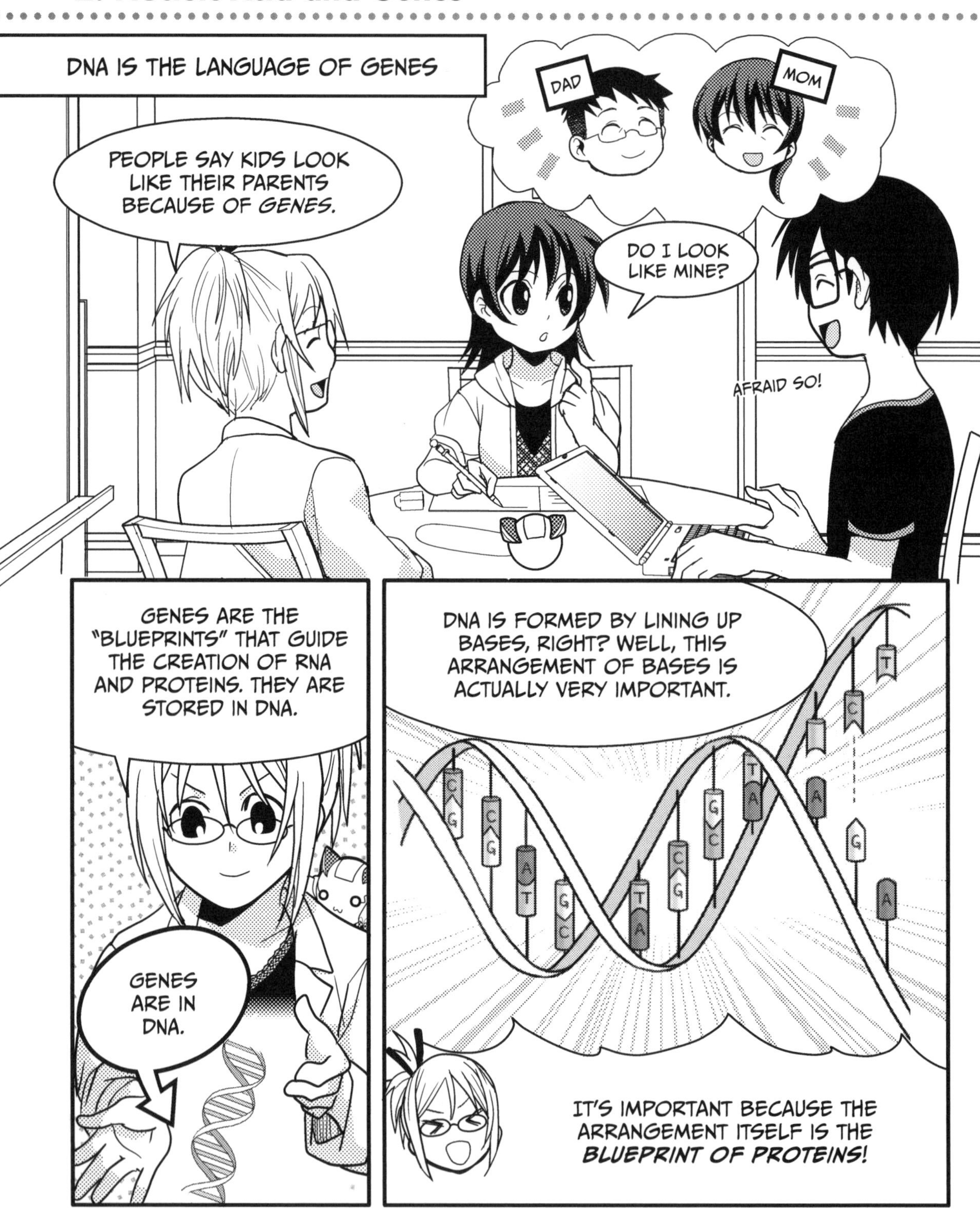

TTCGCGATGCTAGCTATA

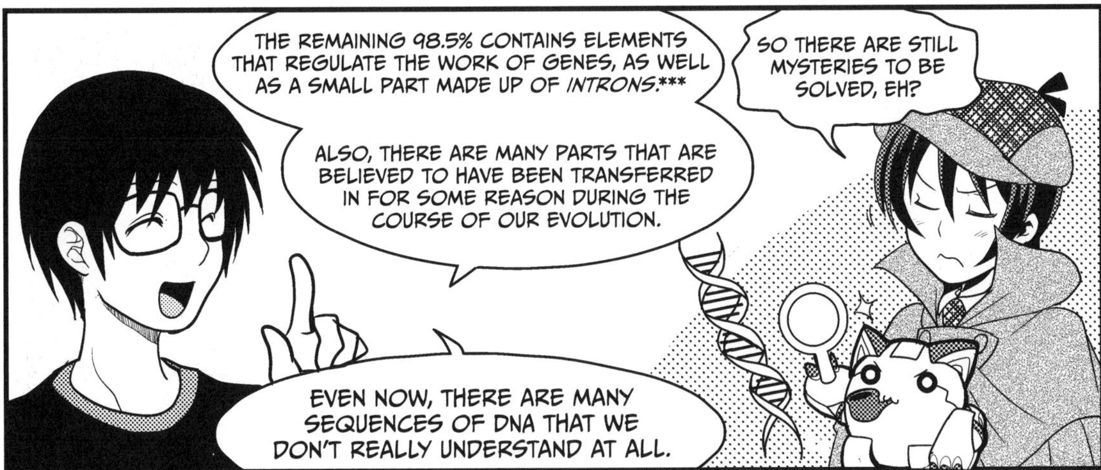

* IN THIS BOOK, WE'LL GENERALLY SAY THAT GENES ARE "WRITTEN" (IN THE FORM OF A BASE SEQUENCE) IN DNA.

** ACTUALLY, THE *EXON* PART OF GENES IS CONSIDERED TO BE 1.5% OF THE WHOLE. (SEE PAGE 227.)

*** WE'LL EXAMINE INTRONS AND EXONS FURTHER ON PAGE 227.

RNA HAS SEVERAL JOBS

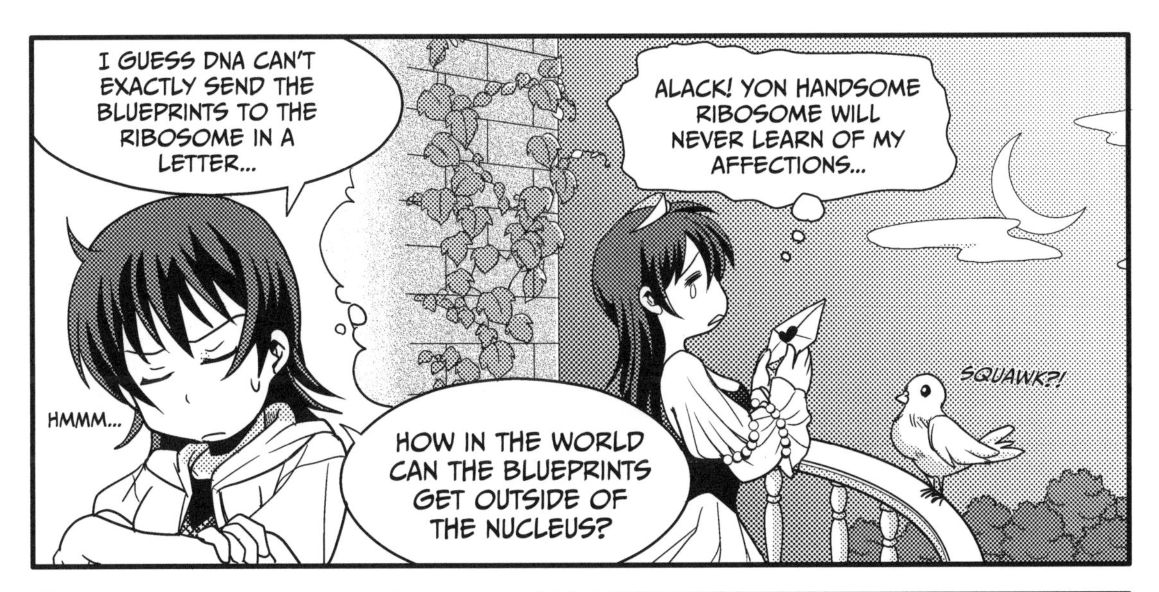
I GUESS DNA CAN'T EXACTLY SEND THE BLUEPRINTS TO THE RIBOSOME IN A LETTER...
ALACK! YON HANDSOME RIBOSOME WILL NEVER LEARN OF MY AFFECTIONS...
SQUAWK?!
HMMM...
HOW IN THE WORLD CAN THE BLUEPRINTS GET OUTSIDE OF THE NUCLEUS?

THAT'S WHERE RNA SAVES THE DAY!
GASP!
YOU DON'T SAY!

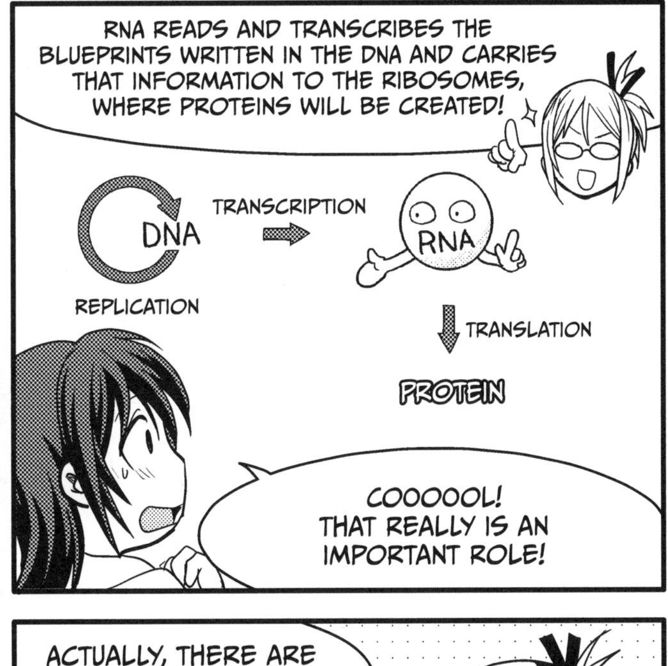
RNA READS AND TRANSCRIBES THE BLUEPRINTS WRITTEN IN THE DNA AND CARRIES THAT INFORMATION TO THE RIBOSOMES, WHERE PROTEINS WILL BE CREATED!
DNA
TRANSCRIPTION
RNA
REPLICATION
TRANSLATION
PROTEIN
COOOOOL! THAT REALLY IS AN IMPORTANT ROLE!

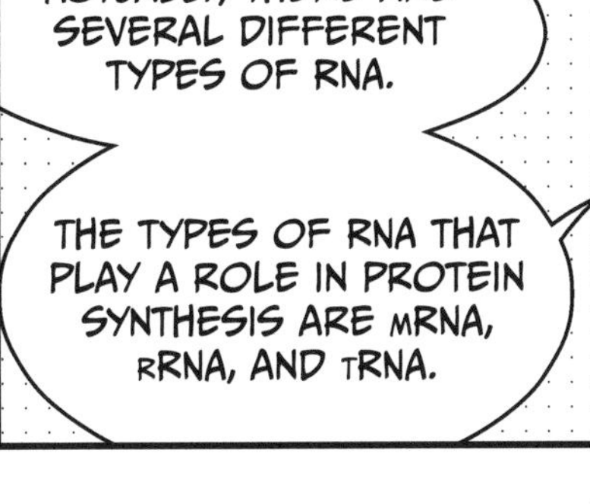
ACTUALLY, THERE ARE SEVERAL DIFFERENT TYPES OF RNA.
THE TYPES OF RNA THAT PLAY A ROLE IN PROTEIN SYNTHESIS ARE mRNA, rRNA, AND tRNA.

mRNA

The enzyme called *RNA polymerase* enables genes that are written in the DNA—that is, their base sequences—to be used as templates to synthesize RNA with complementary base sequences.

This process is called *transcription* (it can also be referred to as RNA synthesis), and the synthesized RNA is called *messenger RNA* (*mRNA*).

For example, take a look at the gene shown in the following figure.

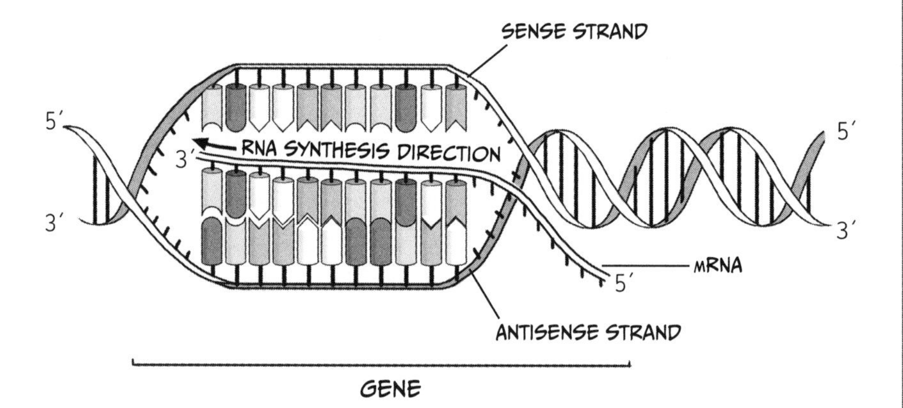

DNA is double stranded, and, for each gene, only one side of the DNA is meaningful (that is, its base sequence becomes the blueprint). This is the *sense strand* for that gene. The other strand, which is complementary to the sense strand, is called the *antisense strand*.

That makes...sense.

Since mRNA is synthesized using the antisense strand as a template, the base sequence of the RNA that's produced is the same as that of the sense strand. However, where the base is T (thymine) in the sense strand, there will be a U (uracil) in the mRNA.

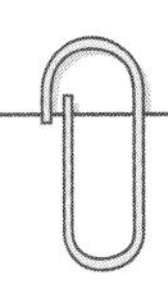

The synthesized mRNA is called *precursor mRNA*. It undergoes a series of chemical reactions and other processing* to become mature mRNA, which is then transported from the nucleus to the cytoplasm and is soon delivered to a ribosome.

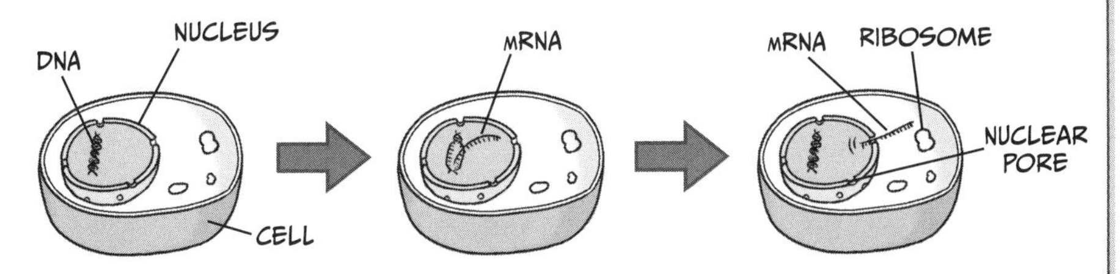

So you can see why we call it a "messenger": it's transporting vital information out of the stronghold of the nucleus.

* This process includes *splicing*, which removes intron parts (see page 227), and the addition of a 5′ "cap" and a poly-A tail.

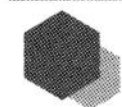

rRNA AND tRNA

A ribosome is like a big protein-making machine consisting of several types of RNA called *ribosomal RNA* (*rRNA*) and several dozen types of ribosomal proteins.

They're not as large as mitochondria or chloroplasts, though.

They may not look like much, but ribosomes are pretty darned impressive!

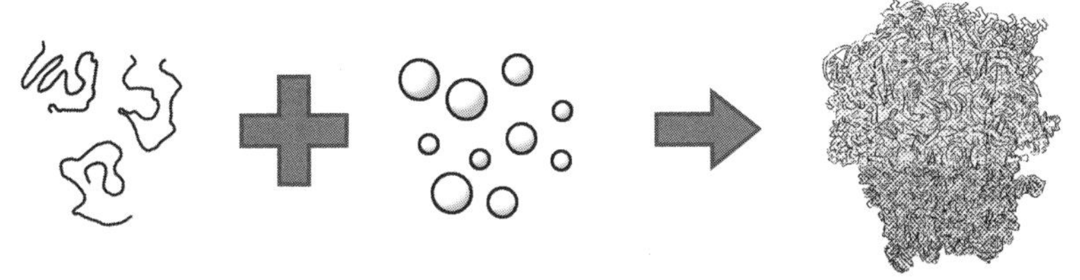

Another type of RNA, which transports amino acids, is called *transfer RNA* (*tRNA*). The tRNA matches the base sequences of the delivered mRNA with the amino acids needed to build the coded proteins.

It should be pretty obvious where transfer RNA gets its name: it **transfers** amino acids.

The mRNA base sequence is a code for an amino acid sequence. More specifically, a three-base sequence forms the code for one amino acid. This three-base sequence is called a *codon*.

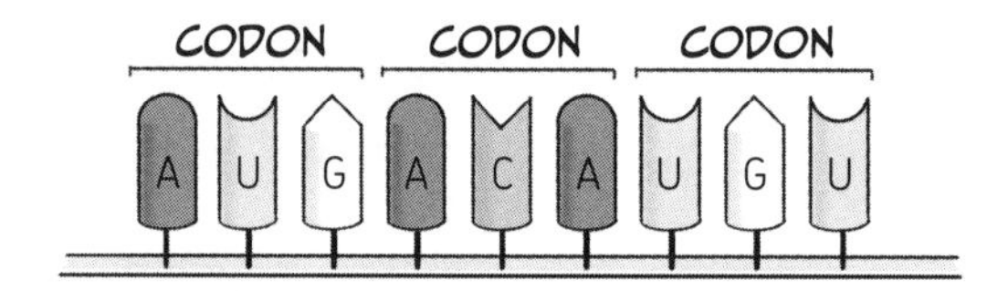

There are unique tRNA molecules for each codon. In other words, only a tRNA with the proper base sequence—called an *anticodon*—that is complementary to the mRNA's codon can adhere at the site of the ribosome.*

The role of rRNA is to connect the amino acids that were transported via tRNA into long chains. Amino acids are connected according to the mRNA base sequence (that is, the codon sequence) to create proteins in accordance with the base sequence (the genes, or "blueprint") of the original DNA.

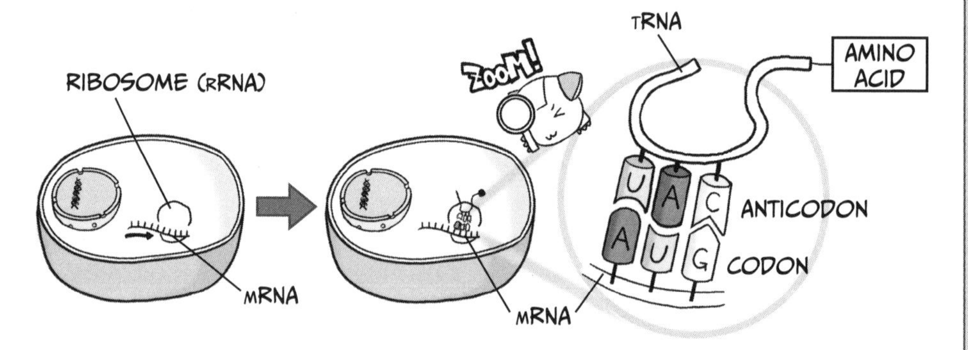

So in other words, genetic information is conveyed like this: DNA → RNA → protein.

* The type of amino acid that is transported is also determined by the anticodon.

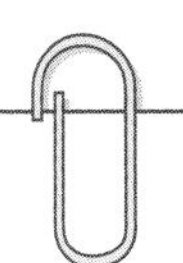

Here's what a single rRNA looks like:

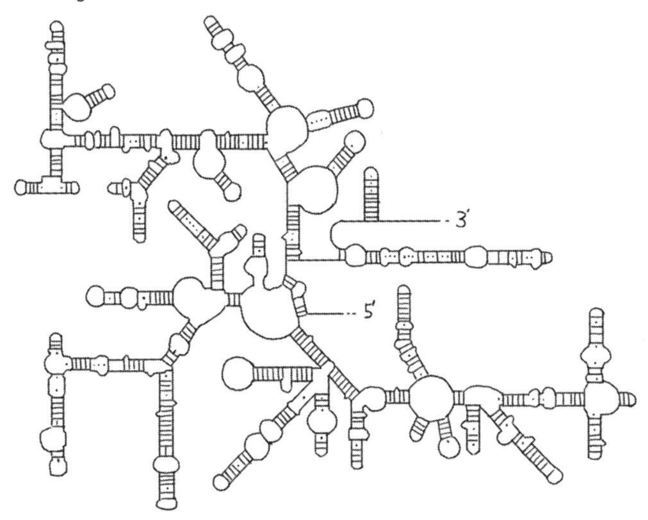

Huh? It looks like a maze.

If you look carefully, you can see that it's actually just **one RNA strand folded in a complex way**. The ends are marked 3′ and 5′.

Oh! You're right!

When an RNA strand folds, its bases pair with each other. In the image above, those pairings are represented by the transverse lines, or "ladder rungs." Base pairing within a single strand is a feature of RNA not found in DNA: **It can fold into a variety of forms according to differences in base sequences**.

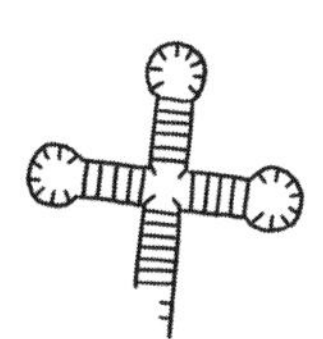

Wow, cool! They look like little line drawings.

RIBOZYMES

To summarize, RNA breaks down quickly, but it can do things that DNA can't. RNA exists both in the nucleus and the cytoplasm, and it can take various forms depending on the base sequence.

It's a very flexible molecule, isn't it?

Definitely. RNA researchers have long believed that RNA probably has other important jobs besides copying genes and interacting with ribosomes.

Ribozymes, which were independently discovered at the beginning of the 1980s by the American microbiologist Thomas Cech and the Canadian molecular biologist Sidney Altman, foretold later developments in RNA research. You learned earlier that one very important job of proteins is their work as enzymes (see page 153), but Cech and Altman actually discovered that RNA has the flexibility to work as an enzyme as well.

Huh!

So they combined the words "RNA (ribonucleic acid)" and "enzyme" to coin the term *ribozyme*.

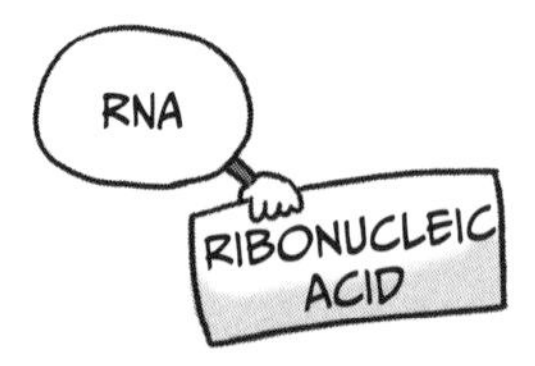

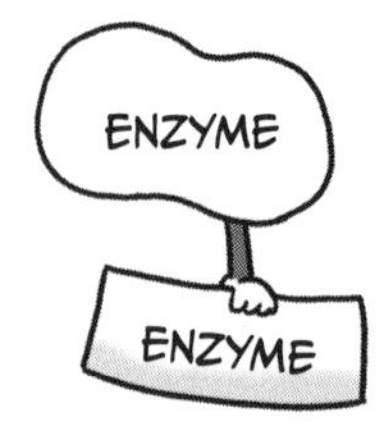

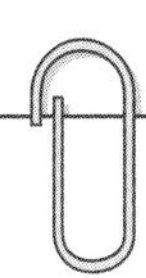

Many of the ribozymes that have been discovered or artificially created so far are involved in building or cleaving RNA or DNA.

Cech discovered that part of RNA itself catalyzes the chemical reaction to remove sections that are meaningless as protein blueprints (introns) and to join the sections that are meaningful (exons). This is referred to as *self-splicing** and is fairly rare.

Usually, splicing involves an enzyme known as *spliceosome*, which is a large complex made of RNA and proteins.

Splicing involves several hydroxyl groups (-OH) of RNA, such as the 2′-OH or the 3′-OH.

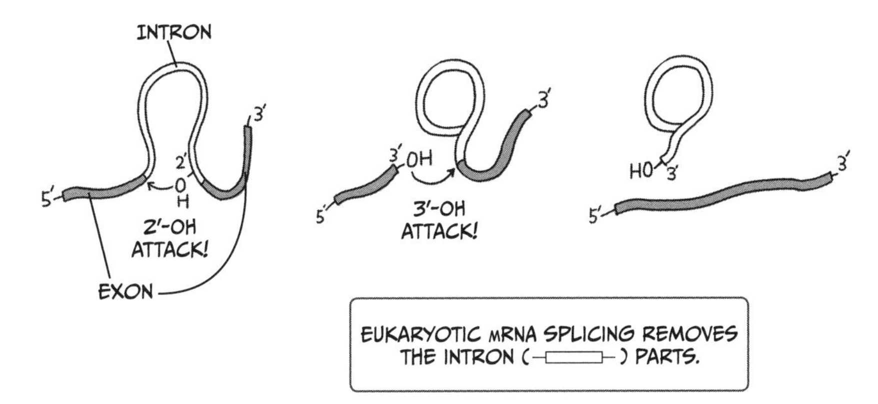

Hey, it made a loop!

The discovery of ribozymes caused RNA to be viewed as a multifaceted molecule that does a variety of tasks, and RNA research advanced rapidly.

As the 21st century began, this research showed that RNA has other various roles and that many different forms of RNA exist in cells. RNA research is truly in full bloom now, and there are high expectations for future research.

* The genes of eukaryotes (like humans) are divided into several exons by non-coding base sequences called introns. The introns must be removed from the mRNA before translation, in a reaction known as splicing.

THE DIRTY JOB OF A BIOCHEMIST

Nowadays—especially in cities—there are fewer and fewer opportunities for exploring nature and really getting your hands dirty. In a world where a mustard-stained tie is the greatest potential tragedy, most people can't even imagine trekking out into the wilderness to really dig through the (often filthy!) wonders of the natural world.

But humans, of course, are products of nature. Long before the miracles of instant spot remover and nectarine-scented hand sanitizer, we lived like other animals, surrounded by the great outdoors.

Although biochemistry is a discipline that attempts to view the incredible phenomena of life through the lens of science, the raw materials of that research are the living organisms themselves—again, products of nature.

Once, during my graduate student days, I went to the mountains in order to collect certain plants that contain large amounts of the proteins that I was researching.

I drove with a younger graduate student to some nearby mountains, and we proceeded along a road that was thickly overgrown with plants. When we got to a place where the car could no longer pass, we continued on foot to find the plant called *Phytolacca Americana* (American pokeweed).

We brought the pokeweed back to the laboratory, washed off the dirt, and used kitchen knives and scissors to cut them up so that we could extract the target proteins from them.

I would also routinely visit a nearby meatpacking plant to obtain an organ called the *thymus* from recently slaughtered cows. (The thymus is usually discarded, so they were happy to give them away free of charge.) I would use scissors to cut up the organ and store the little pieces in the laboratory's refrigerator as experimental samples. This was necessary to extract the research target protein (DNA polymerase).

In this way, biochemistry originally developed based on a methodology of extracting (isolating, purifying, etc.) chemical substances from organic raw materials and checking their chemical properties.

In contrast, molecular biology is a discipline that tries to elucidate the phenomena of life by using *biopolymers* such as DNA and proteins.

Simply speaking, if molecular biology just deals with DNA and RNA or prepares an environment for artificially creating proteins (by using *E. coli*, for example), it is unnecessary to use flesh-and-blood, organic raw materials such as a cow's thymus or plant materials.

In a manner of speaking, the data obtained in these artificial lab environments is "digital." Molecular biology is practiced using high-tech equipment and cutting-edge technologies, and molecular biologists are rarely caked in mud or smeared with animal blood...

As a result, some biochemists refer to their own research as "dirty work." I think that when they do this, it's somehow self-deprecating.

However, it is an undeniable fact that the accumulation of knowledge from that kind of dirty work built the foundation of molecular biology in the first place. Although many young molecular biologists have never once used flesh-and-blood, organic materials other than *E. coli*, cell cultures, or experimental animals, biochemistry and molecular biology have long been bound together and remain so even today. This fact should never be forgotten.

EARLY BIOCHEMISTRY AND MOLECULAR BIOLOGY

In 1897, German biochemist Eduard Buchner made the revolutionary discovery that *fermentation* occurred in an extract created from yeast cells, which contained yeast protein but no living yeast cells.

Prior to that, people believed that the chemical reaction called fermentation, which is characteristic of living organisms, could not happen without the presence of living cells. However, this idea was utterly obliterated by Buchner's discovery.

Because of this, the theory of *vitalism*, in which the phenomena of life occur only because of a characteristic force (spirit or life energy) of living organisms, practically disappeared, and the foundation was built for research on chemical reactions occurring in living organisms in test tubes. In other words, biochemistry was born.

Since Buchner found that an actual living organism need not be present, it would not be an exaggeration to say that this was the discovery that paved the way for the arrival of molecular biology.

As a clearer understanding of the chemistry of life developed, it became apparent that certain mechanisms common in all living organisms form the basic foundation of life. Some of these mechanisms, for example, are that DNA is used as the genetic language, that the basic theory (central dogma) behind the reading of genes to create proteins is consistent, and that the same proteins often do the same kinds of work.

Because these mechanisms are universal among living organisms, it was important to develop methodologies for studying DNA, which serves as the blueprint for proteins, and for elucidating the function of proteins.

DEVELOPMENT OF RECOMBINANT DNA TECHNIQUES

In 1972, American molecular biologist Paul Berg performed the first successful recombinant DNA experiment in the world by artificially manipulating DNA in a test tube to create a DNA sequence that did not exist in the natural world.

A method for easily decoding the base sequences of DNA was developed by English biochemist Frederick Sanger in 1977, and a method for copying (or *amplifying*) DNA was developed by American molecular biologist Kary Mullis in 1985. After these discoveries, recombinant DNA experimental techniques advanced rapidly.

Recombinant DNA techniques developed after it had been clearly demonstrated that DNA is the genetic language (or, in other words, that the information referred to as genes is written in the form of base sequences in DNA). Molecular biologists imagined that if they only had the DNA base sequences and if they only created an environment in which proteins could be created from those base sequences, they could elucidate the phenomena of life, which can be thought of as the totality of chemical reactions in which proteins are involved.

For example, if genes from an external source are introduced into a simple, easy-to-manipulate organism for which these mechanisms are well understood (such as *E. coli*) to create proteins within that organism, a large quantity of proteins can be obtained all at once without ever having to perform any more "dirty work."

Generally speaking, this idea can be expressed as: Explain the DNA sequences! Explain life!

Molecular biology has developed with this aim. The methodology used to achieve this was based on recombinant DNA techniques.

RETURNING TO BIOCHEMISTRY

After the Human Genome Project (an international cooperative research project aiming to sequence the genetic information of humans) was completed in 2003, researchers turned their attention from DNA back towards proteins and RNA.

The *post-genome era*, or *post-sequencing era*, arrived.

No matter how much DNA steals the limelight, no matter how its handling techniques develop, and no matter what anyone says, when life phenomena are viewed as a collection of "chemical reactions," the only things at work there are proteins and RNA.

This is because, even if all of the base sequences of human DNA (that is, the entire genome) are known, the data is meaningless unless the roles of the proteins and RNA that are made from DNA are understood.

Currently, if the amino acid sequences of many proteins are known and the role they play is also understood, then the work of unknown proteins can be predicted to a certain degree based on only amino acid information.

However, the role of unknown proteins must be verified by using biochemical techniques. No matter how many molecular biological techniques (such as DNA recombination) are used to research the activity of proteins to elucidate their roles, the question of whether or not those proteins really do those jobs inside natural cells will remain unanswered. In a manner of speaking, this is the same as the idiomatic expression "you can't see the forest for the trees." As long as the research target is a biological substance, biochemistry remains an absolutely necessary and important academic discipline.

THE ORIGIN OF THE CELL

No doubt you've heard the phrase "origin of life." We're not going to tackle the intimidating question of what life actually is here. Instead, we're concerned with the first living organisms or, in other words, "the origin of cells."

So how did cells originate on Earth?

By now, you should understand that a cell is the location of many vital chemical reactions in living organisms and that cells can divide and multiply. These are two major traits that define life.

Chemical reactions that transform one substance into another substance, which are performed inside of cells, and networks of many of these chemical reactions, which occur across many cells, are collectively referred to as *metabolism*.

Cells must constantly undergo metabolism in order to produce the energy required to maintain their organization and to reproduce.

In addition to metabolism, reproduction, also called *self-replication*, is a major characteristic of living organisms. Unicellular organisms can often simply replicate their cellular contents and split in two; this process is called *binary fission*. For multicellular organisms, reproduction is more complex and involves specialized reproductive cells, which initiate the development of offspring.

(Although the way in which cells replicate themselves can certainly be called self-replication, the expression is a little awkward. Let's just call it plain old "replication" from now on.)

Cells perform both metabolism and replication, both of which are extremely important when considering the origin of life. But which came first, metabolism or replication?

This is one of the most difficult questions confronting scientists who are conducting research concerning the origin of life. When the "pouch wrapped in a membrane" (the cell) originated, what had been occurring inside that enclosure? Some scholars think that the "pouch" had been performing metabolism after many diverse molecules had collected together. Some time later, the "pouch" absorbed molecules capable of causing replication, so the entire assembly became able to divide and multiply. This idea is known as *metabolism first*.

Other scholars think that the original "pouch" had enclosed replicating molecules and had been dividing. Later, it became able to perform metabolism, and it eventually obtained a more advanced replication method, causing it to evolve into a cell. This idea is known as *replication first*.

Many other scholars think that posing the question of "which came first" is nonsense and that metabolism and replication coevolved cooperatively.

In any case, the biochemical process of metabolism is a very large-scale phenomenon.

4. Conducting Biochemistry Experiments

It was mentioned on page 230 that the functions of proteins can be verified by using biochemical techniques, but what kinds of experiments do biochemists perform?

There are various experimental methods depending on the field of research, and not each and every one will be presented here. Instead, I will introduce several experimental methods that I have used myself.

COLUMN CHROMATOGRAPHY

Column chromatography is an experimental method for separating substances that have the same property from a mixture of substances. For example, we can collect just proteins with certain properties from the liquid extract of American pokeweed, which was introduced earlier, or from the liquid obtained after a cow thymus was blended in a juicer. Special *resins* are packed in a long, narrow tube made of glass or some other material. Some substances adhere to those resins and are trapped, allowing for the collection of only those substances that do not adhere.

Various types of chromatography, such as ion exchange chromatography, gel filtration chromatography, and affinity chromatography, can be used, depending on the type of resin or target protein. As an example, here we will look at a method for purifying the enzyme DNA polymerase α from the thymus of a calf.

As shown below, the cells are smashed by grinding up the calf thymus, and a solution with a high salt (sodium chloride) concentration is used to extract proteins and other molecules. This extract (the liquid in the flask) passes through the large glass tube (called the *column*) in which the *ion exchange resin* has been packed. The proteins are broadly divided into those that are trapped in the column and those that pass through the ion exchange resin.

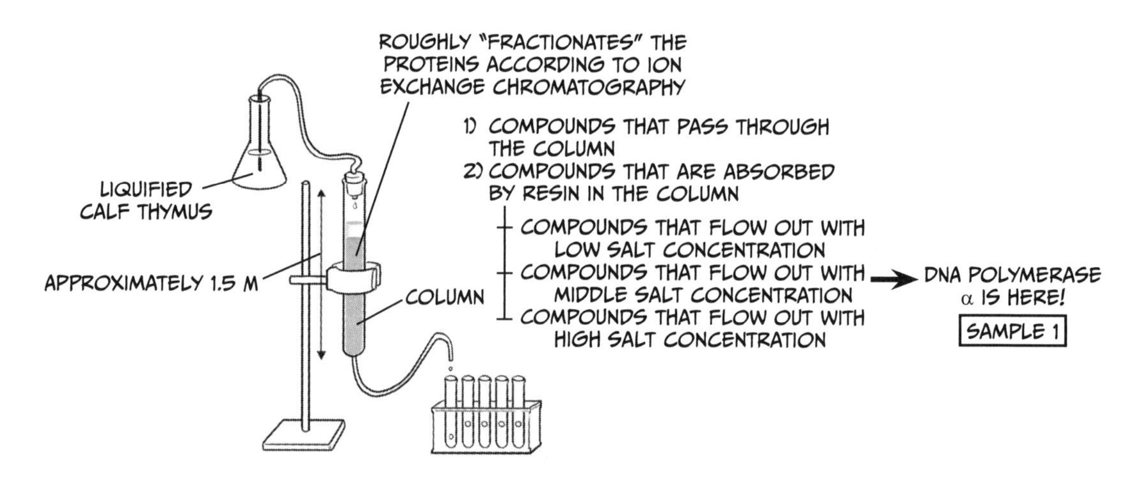

The substances that pass through the ion exchange resin are collected in test tubes, but the substances that are absorbed are separated from the resin and collected in test tubes later by adding a liquid with a higher salt concentration. DNA polymerase α adheres to the column and can be retrieved by adding a liquid with a salt concentration of 0.5 M (moles per liter). (This is "sample 1.")

The figure below shows the method of purifying DNA polymerase α from this "sample 1."

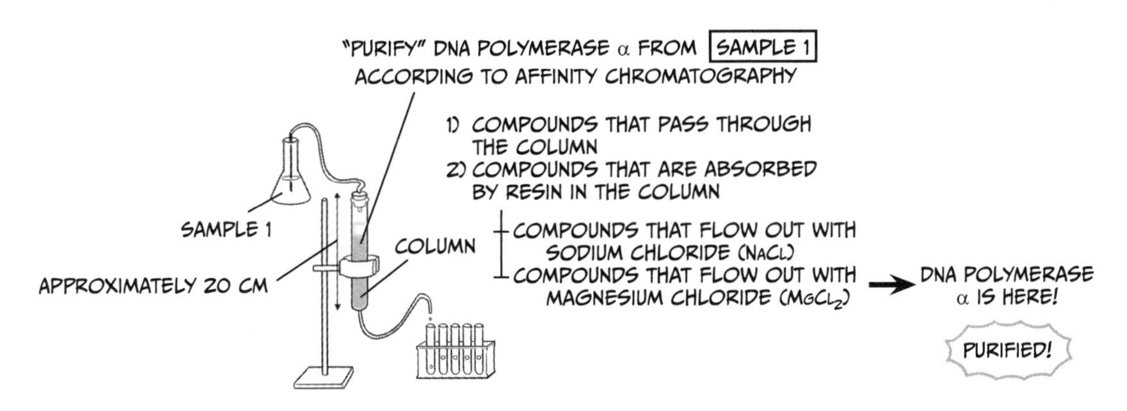

This technique, called *affinity chromatography*, uses a small glass tube packed with a resin combined with an antibody (a type of protein that's created by the immune system) that can only bond with DNA polymerase α. When sample 1 is passed directly through this resin, it's divided into two parts—one that is absorbed and one that passes through. DNA polymerase α, which is absorbed, can be retrieved by running a solution with an extremely high concentration (3.2 M) of magnesium chloride through the resin. Since the part that's retrieved here is almost 100% DNA polymerase α, it has been "purified" at this time.

In this way, DNA polymerase α can be efficiently purified by using a combination of ion exchange chromatography and affinity chromatography.

ELECTROPHORESIS AND A WESTERN BLOT

This is an experimental method for isolating a specific protein, recognizing what type of protein is in a sample, or checking the size of a target protein. *Electrophoresis* is a method in which a sample is "loaded" on top of a thin gelatinous slab (gel) and a current passes through it to cause the sample to move through the gel. SDS-polyacrylamide gel electrophoresis (left side of the following figure), which separates proteins based on molecular size, is commonly used. After the proteins are separated, they are detected by reacting them with a special reagent, such as a dye or a fluorescent marker.

A *western blot* (right side of the following figure) is a method of transferring the proteins that are in the gel onto a thin membrane, while retaining the positions that they were in after they were separated, and then detecting a specific protein on the membrane by adding an "antibody" that only reacts with that protein. *Lectin blotting*, which is described next, is an application of a western blot.

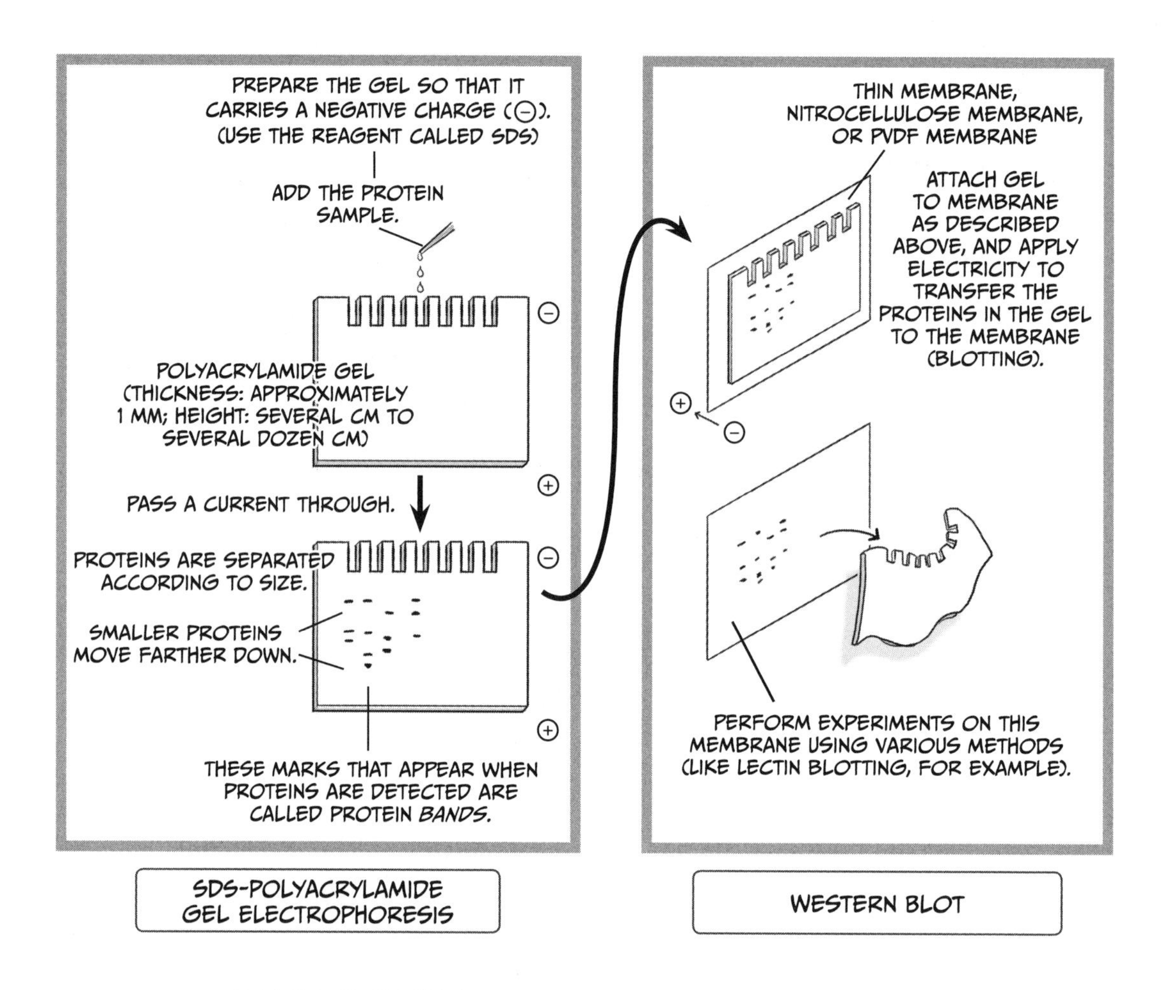

LECTIN BLOTTING

Lectins are proteins that can bond specifically to certain sugar chains. Because lectins will bond differently according to the type of sugar chain, lectins can be used to identify the type of sugar chain that is bonded to a protein. A method similar to a western blot can be used after proteins are transferred onto a membrane. Various lectins are mixed with the proteins, and only the lectins that reacted are detected. This lets us identify the types of sugar chains that exist in the proteins that were transferred onto the membrane. This is called lectin blotting.

The following figure shows an experiment in which a lectin called wheat germ agglutinin (WGA) identifies a sugar chain to which the saccharide called *N*-acetylglucosamine (GlcNAc) is attached.

In this lectin blot performed for the rough protein fraction of the starfish oocyte, it is apparent that two large bands are glowing brightly. WGA was the lectin used here.

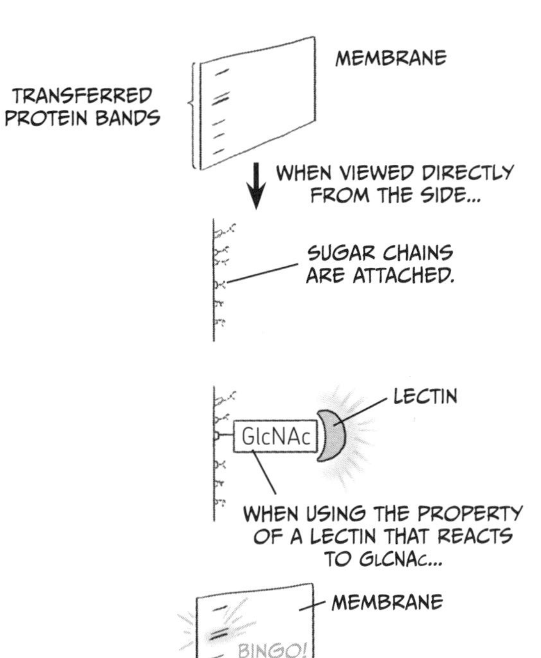

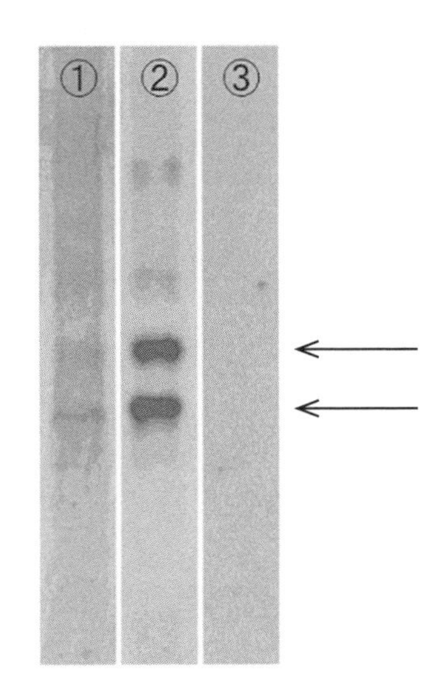

① ALL PROTEINS ARE STAINED WITH CBB.
② ONLY PROTEINS POSSESSING SUGAR CHAINS WITH GLCNAC ATTACHED ARE STAINED BECAUSE OF WGA (ARROWS).
③ THESE PROTEINS ARE STAINED WITH CONCANAVALIN A, WHICH IS ANOTHER LECTIN.

(PHOTOS PROVIDED BY: MITSUTAKA OGAWA, NAGAHAMA INSTITUTE OF BIO-SCIENCE AND TECHNOLOGY)

CENTRIFUGATION

Like column chromatography, *centrifugation* is an experimental method that is performed to separate organelles, proteins, and other biological molecules that have the same density from a mixture of substances with differing densities. The sample solution is placed in a test tube, and the test tube then spins around the centrifuge axis at a very high speed. To handle small molecules, like proteins, ultracentrifugation may also be performed, which spins the test tube at tens of thousands of revolutions per minute. DNA and other polymers can also be isolated in this way.

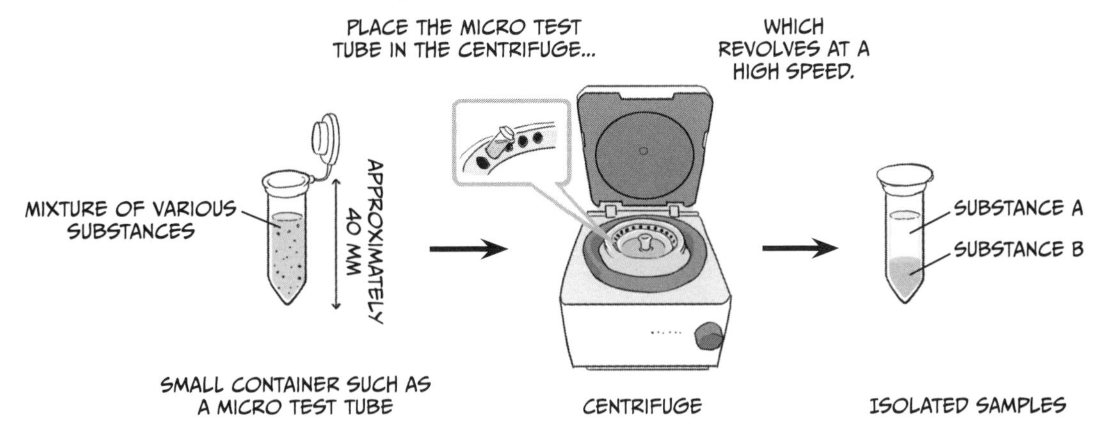

ENZYME REACTION MEASUREMENT

There are various methods of measuring enzyme activity, such as using radioisotopes to measure the amount of a reaction product produced or using a change in color to measure the reaction product formed when a substrate is acted on by an enzyme.

I will explain a method of using radioisotopes to measure the enzyme activity of DNA polymerase and a method of using a color reaction to measure the enzyme activity of α-amylase.

DNA POLYMERASE ACTIVITY MEASUREMENT METHOD

First, place the following in a mirco test tube: the solution for measuring activity (with pH optimized and buffered for DNA polymerase), DNA polymerase, the DNA that is to be the template, the nucleotide that is to be the raw material, and the magnesium chloride co-factor. Add the nucleotide that contains the radioisotope, and let it react at 37° C for a fixed interval.

When this is done, the nucleotides that contain the radioisotopes are captured in the new DNA strands that have been synthesized by the DNA polymerase. The unreacted nucleotides are removed, and the synthesized DNA strands are placed in a small radioisotope bottle for measurement in a device called a *liquid scintillation counter*, which measures radioisotopes. Since higher enzyme activity means more radioisotopes were captured in the DNA, higher numeric values indicate higher enzyme activity.

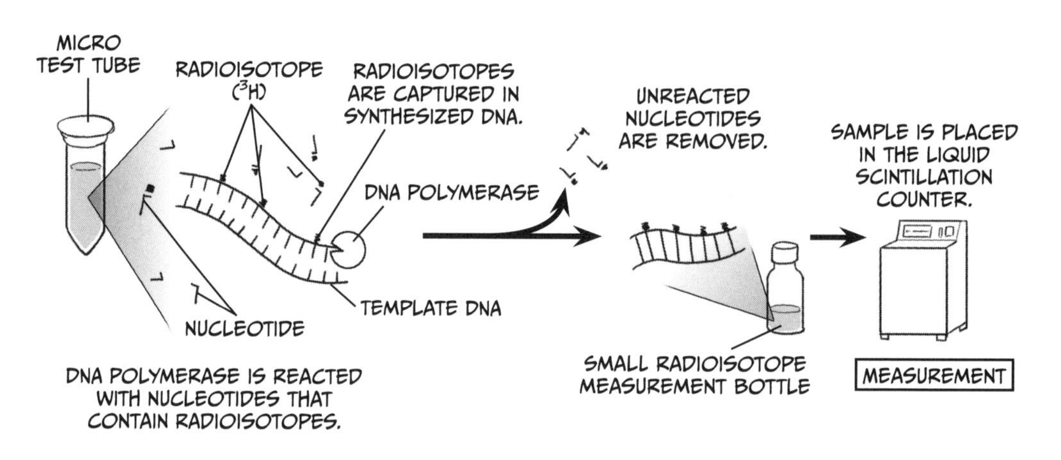

DNA POLYMERASE ACTIVITY MEASUREMENT METHOD

α-AMYLASE ACTIVITY MEASUREMENT METHOD

An α-amylase solution (such as saliva) is added to a solution in which starch has been dissolved inside a test tube. If an iodine solution is immediately added to this, before any of the starch has been broken down, the starch will react with the iodine and produce a blue-violet color. However, as time passes, the starch is broken down by the α-amylase, and the color steadily changes (blue-violet → violet → red → orange → pale orange). Eventually, when all the starch has been broken down, the solution will become colorless. The enzyme activity of α-amylase can be measured by using a spectrophotometer to quantify the appearance of its color as a numeric value.

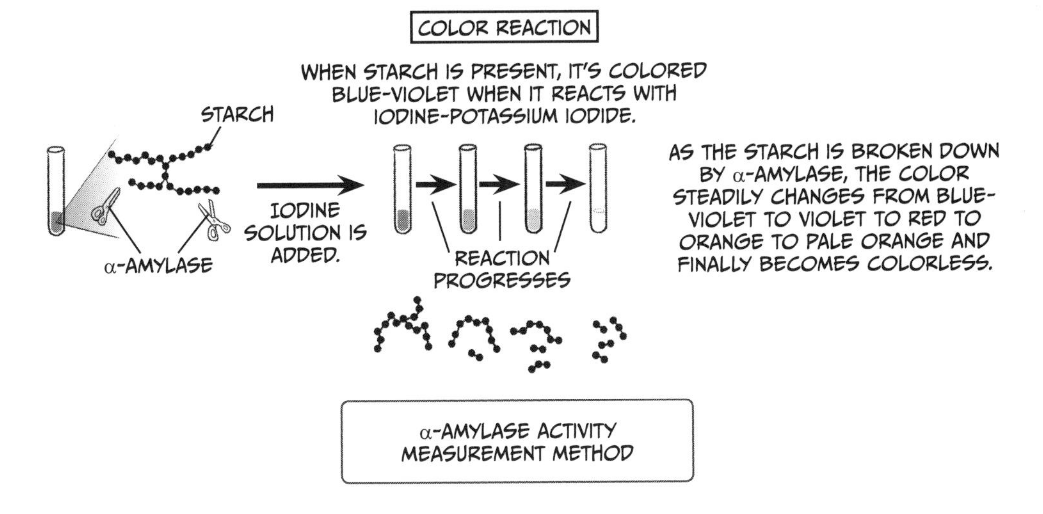

α-AMYLASE ACTIVITY MEASUREMENT METHOD

* PROJECTION ROOM
スクリーン室*
WELL, THAT'S THE END OF OUR LESSONS!
WHAM
YOU TWO DID A GREAT JOB!
IT'S ALL THANKS TO YOU, PROFESSOR!
UM...BY THE WAY, PROFESSOR...
NOW THAT WE'VE FINISHED OUR LAST LESSON...
THE TIME HAS COME!
キャー
TEACH ME THE ULTIMATE SECRETS OF DIETING!

?
SILLY GIRL—YOU'VE HAD THE "SECRETS" OF DIETING THIS ENTIRE TIME!
WHA-?
LOOK...
生化学 レポート
食べすぎると太っちゃうのはどうして？
TAP
IT'S ALL RIGHT HERE IN YOUR REPORT.
THIS GRAPH IS THE KEY!
INGESTED ENERGY
EXPENDED ENERGY
ENERGY STORED AS FAT
EAT MODERATELY AND EXERCISE FREQUENTLY!
ALTHOUGH THERE ARE ENDLESS FAD DIETS OUT THERE, IT ALL REALLY BOILS DOWN TO THAT!
IF YOU EXPEND MORE CALORIES THAN YOU INGEST, YOU'LL LOSE WEIGHT.

AND MAKE SURE YOU GET A GOOD BALANCE OF NUTRIENTS, OF COURSE!
YOU'VE LEARNED THAT PROTEINS, SACCHARIDES, AND LIPIDS ARE ALL IMPORTANT TO YOUR BODY. YOU MUST REALIZE BY NOW THAT DIETING BY STARVING YOURSELF IS COMPLETE NONSENSE, RIGHT?
RIGHT, KUMI? RIGHT?!
STAGGER STAGGER
GUUUUUUUH
ずーーーん
...OH DEAR.
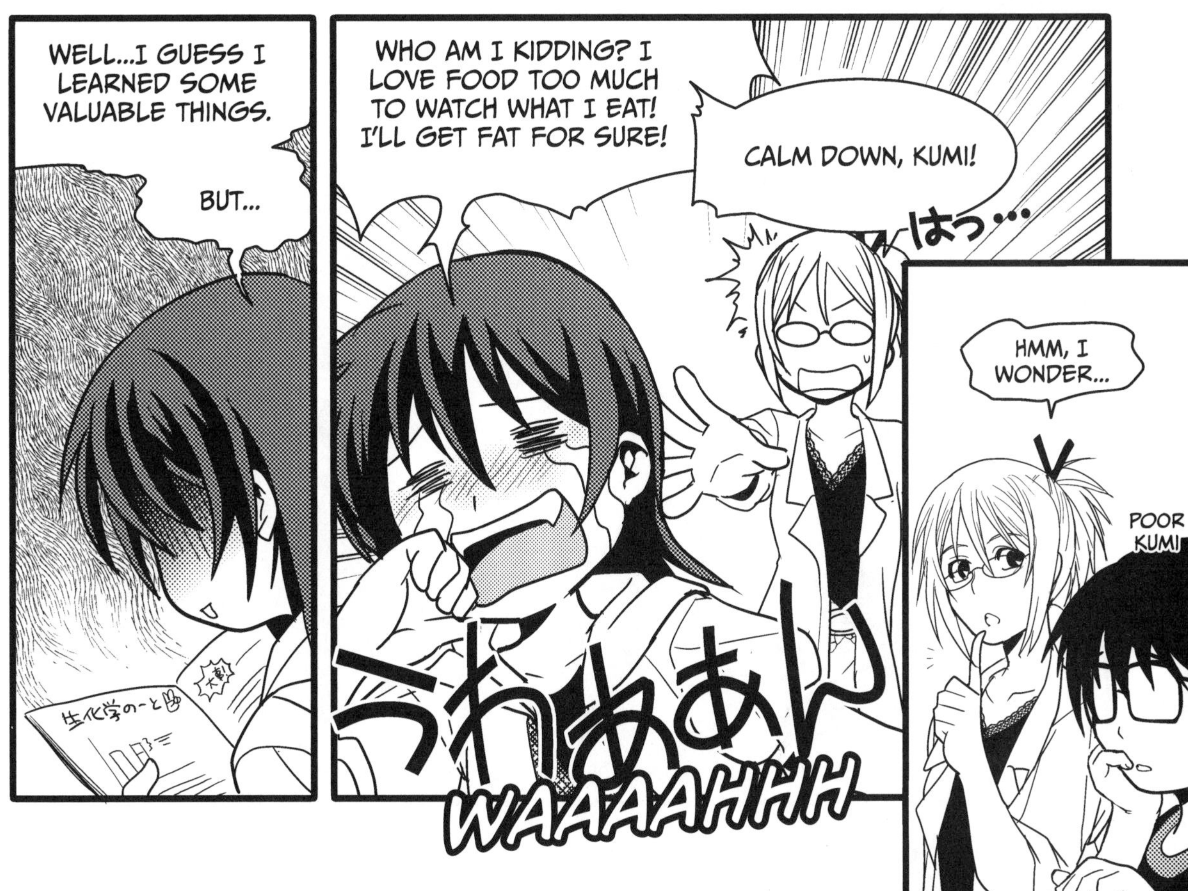
WELL...I GUESS I LEARNED SOME VALUABLE THINGS.
BUT...
生化学のーと
WHO AM I KIDDING? I LOVE FOOD TOO MUCH TO WATCH WHAT I EAT! I'LL GET FAT FOR SURE!
うわああん
WAAAAAHHH
CALM DOWN, KUMI!
はっ...
HMM, I WONDER...
POOR KUMI

DOOMED!
I'LL EAT CAKE AND PIZZA UNTIL I'M HUGE!
I'LL TURN INTO A BLOATED, SNACK-SCARFING MANATEE!
KUMI, YOU'RE BEING RIDICULOUS. YOU'VE GOT A LONG WAY TO GO BEFORE YOU'RE AS BIG AS A MANATEE.
IF ANYTHING, YOU'D BE MORE LIKE A MIDSIZED SEA LION...
HEH HEH HEH
...OR MAYBE A SMALL BEAR PREPARING FOR HIBERNATION, COVERED IN WINTER FAT.
SQUIRM
SQUIRM
HEY!
FATTY!
WHAT?
HA HA HA HA
BUT YOU KNOW... IF YOU THINK ABOUT IT, CHUBBY ANIMALS ARE ALMOST ALWAYS THE CUTEST, SO DON'T WORRY!
HEY, PROFESSOR...
SOB
SOB
SOB

GIVE KUMI A BREAK!
CAN'T YOU SEE SHE'S TOO SENSITIVE TO STAND UP TO THIS KIND OF TREATMENT?
OH YEAH? IF YOU HAVE SOMETHING TO SAY, YOU'D BETTER SPILL IT!
SNIFF
I WILL!
FIRST OF ALL, I DON'T THINK KUMI'S FAT AT ALL.
GOAL: LOSE 5 LBS!
DOWN WITH THE POUNDS!
BUT EVEN IF SHE *WAS* FAT, SHE'D STILL BE BEAUTIFUL!

SHE'S BEAUTIFUL WHEN SHE'S STUFFING HER FACE...
YUM
HMMM...
SHE'S BEAUTIFUL WHEN SHE'S PORING OVER HER STUDIES...
AND SHE'S EXTRA BEAUTIFUL IN HER BATHING SUIT!
IN FACT...
SHE'S ONE OF THE MOST BEAUTIFUL GIRLS I'VE EVER MET! I'D EVEN GO AS FAR AS SAYING THAT...
KUMI IS A TOTAL DREAMBOAT!

WELL DUUUUH! YOU THINK I DON'T KNOW THAT?
HUH?
KUMI IS A CUTIE PATOOTIE!
...AND THE "FAT LITTLE BEAR"...
BUT...THAT "SEA LION" CRACK...
かああ
NE-
NEMOTO?

JUST A LITTLE REVERSE PSYCHOLOGY!
WINK
WORKED LIKE A CHARM, TOO. IT'S ABOUT TIME NEMOTO TOLD YOU HOW HE FELT! ♪
YES, THIS LESSON WAS A HUGE SUCCESS, IF I DO SAY SO MYSELF.
A TRAP...
KUMI, YOU STUDIED HARD AND LEARNED SOME IMPORTANT THINGS ABOUT YOUR BODY, RIGHT?
DEFINITELY. NO MORE FAD DIETS FOR ME.
AND YOU ALSO LEARNED SOME *OTHER* IMPORTANT INFORMATION...HEHE.
カァアア~
BLUSH

SO NOW THAT WE'RE THROUGH WITH OUR LESSONS, HOW ABOUT WE GO GRAB A BITE TO EAT?
YES!
NEMOTO'S TREAT, OF COURSE.
BUT...BUT...I'M STILL TOTALLY BROKE!
OH, I'M SURE A SMART KID LIKE YOU CAN FIGURE SOMETHING OUT.
I'LL GO GET READY!
...
アセ アセ
UM...NEMOTO...

THANKS! FOR EVERYTHING!
!
AND NOW...
LET'S GO STUFF OUR FACES!

INDEX

D

E

M

N

O

P

NOTES

NOTES

NOTES

NOTES

ABOUT THE AUTHOR

Masaharu Takemura, PhD, is currently an Associate Professor at the Tokyo University of Science. His specialties are molecular biology and biology education.

PRODUCTION TEAM FOR THE JAPANESE EDITION

Production: Office Sawa
Email: *office-sawa@sn.main.jp*

Established in 2006, Office Sawa has produced numerous practical documents and advertisements in the fields of medicine, personal computers, and education. Office Sawa specializes in manuals, reference books, or sales promotional materials that frequently use instructional text and manga.

Scenario: Sawako Sawada

Artist: Kikuyaro

DTP: Office Sawa

HOW THIS BOOK WAS MADE

The *Manga Guide* series is a co-publication of No Starch Press and Ohmsha, Ltd. of Tokyo, Japan, one of Japan's oldest and most respected scientific and technical book publishers. Each title in the best-selling *Manga Guide* series is the product of the combined work of a manga illustrator, scenario writer, and expert scientist or mathematician. Once each title is translated into English, we rewrite and edit the translation as necessary and have an expert review each volume. The result is the English version you hold in your hands.

MORE MANGA GUIDES

Find more *Manga Guides* at your favorite bookstore, and learn more about the series at *http://www.nostarch.com/manga*.

UPDATES

Visit *http://www.nostarch.com/mg_biochem.htm* for updates, errata, and other information.

The Manga Guide to Biochemistry was laid out in Adobe InDesign. The fonts are CCMeanwhile and Chevin.

This book was printed and bound at Sheridan Books, Inc. in Chelsea, Michigan. The paper is 60# Finch Offset, which is certified by the Forest Stewardship Council (FSC).